Information and Communication Technology

for AS Level

Julian Mott and Anne Leeming

Hodder & Stoughton

A MEMBER OF THE HODDER HEADLINE GROUP

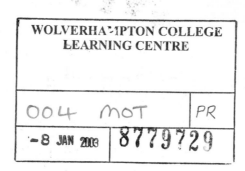
Orders: please contact Bookpoint Ltd, 130 Milton Park, Abingdon, Oxon OX14 4SB. Telephone: (44) 01253 827720, Fax: (44) 01235 400454. Lines are open from 9.00 – 6.00, Monday to Saturday, with a 24 hour message answering service. You can also order through our website www.hodderheadline.co.uk.

British Library Cataloguing in Publication Data
A catalogue record for this title is available from the British Library

ISBN 0 340 80427 0

First published 2002
Impression number   10 9 8 7 6 5 4 3 2
Year   2005 2004 2003 2002

Copyright© 2002 Julian Mott and Anne Leeming

Typeset by Pantek Arts Ltd, Maidstone, Kent.
Printed in Great Britain for Hodder & Stoughton Educational, a division of Hodder Headline Ltd, 338 Euston Road, London NW1 3BH by J.W. Arrowsmiths Ltd, Bristol.

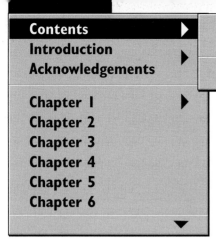

Contents

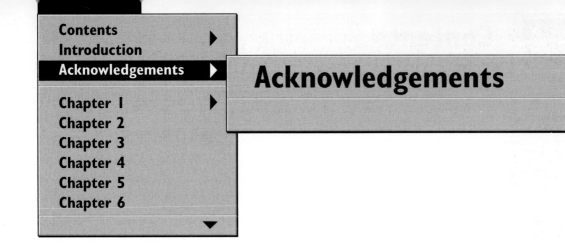

Acknowledgements

Thanks go to Paul Amey of Peter Symonds' College, who produced most of the graphics for this book. He demonstrated cheerful good humour whenever last minuts changes were requested.

Thanks also to Graham Bradshaw and Richard Carr, both ICT teachers at Peter Symonds', who gave up time to review the content of the chapters. They gave vital feedback, ideas for change and valuable material that has been included in the book.

Thanks also to Nicole Stevens and Paul Reed.

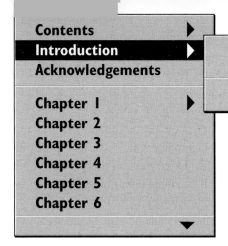

Introduction

This book is designed to cover the examinable modules of the AQA AS Level Information and Communication Technology specification (5521). These modules are also the first two modules of the A Level Information and Communication Technology specification (6521).

The book follows the specification for AQA board very closely but will also be of use to students preparing for other boards at AS level. This book is also suitable for students studying for AVCE ICT.

This book has been substantially re-written from our second edition and it fully reflects the changes in the AQA specification. We have listened to comments from both students and teachers and have included more activities in the chapters as well as providing greater detail and more case studies throughout.

The first ten chapters cover the first module ICT1. The next eleven chapters cover the second module ICT2.

Brief answers are provided to questions. Students should be aware that all examination answers should be in sentences. Questions using words like 'explain' and 'describe' require more detail in the answer.

The AQA AS level ICT specification also includes a coursework module, ICT3. Students studying this module are advised to read *Spreadsheet Projects in Excel for Advanced Level* by Julian Mott and Ian Rendell published by Hodder and Stoughton. ISBN 0-340-80007-0.

Information Communication Technology is a subject that is always changing; we have tried to incorporate many new developments into this book. Students are advised to keep up to date with developments; a good way to do this is through accessing some of the many newspaper and magazine websites shown below.

http://news.ft.com/home/uk/
http://www.cw360.com
http://www.guardian.co.uk/
http://www.computing.co.uk
http://www.independent.co.uk/
http://webserv.vnunet.com/news
http://www.telegraph.co.uk/

Some other useful sites include;

http://www.silicon.com/

http://www.howstuffworks.com

http://whatis.techtarget.com/

http://www.bcs.org.uk (British Computer Society)

For information on the specification and examinations visit AQA's site at:

http://www.aqa.org.uk

For further information on the legislation relating to ICT visit the open government site:

http://open.gov.uk

ASK JEEVES provides an excellent search engine:

http://ask.co.uk

Remember that ICT is all around you; the more you relate what you learn to the real world the easier it will be to learn and answer examination questions. Keep your eyes open wherever you are: in a shop, booking a ticket for a cinema or a train journey, gaining information in a museum or buying a lottery ticket. Keep asking yourself how ICT is involved. Parents, neighbours or family friends may have jobs or take part in leisure activities that involve the use of ICT. Talk to them about topics covered in the book that relate to what they do.

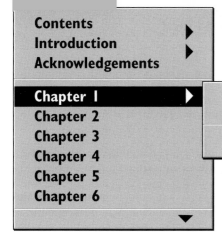

Knowledge, information and data

Information and data

Raw facts and figures that are recorded are **data**. Transactions and events can also be data. A patient's date of birth is a fact; depositing £25 into your bank account is a transaction; a car driving over a pressure pad when leaving a car park is an event.

Facts and figures can be processed to create **information**. Information must have a *context*, which makes it understandable and useful.

For example:

A bank account number is **33 145 121**. This number has no context. It is *data*.

When the account number **33 145 121** is entered into the bank's computer, the name, address and bank balance of the account holder are displayed. The data has been given a context. It is *information*.

It is the context that determines the meaning of information. **33 145 121** might be a bank account number, a bar code on a product, a computer password or the population of California.

Knowledge means applying and interpreting information to make decisions. Knowledge is a set of concepts, rules or procedures that are used to create, collect, store or share information.

The bank can interpret information such as the name, address and bank balance of the account holder to decide whether to write to the account holder because their account is overdrawn. This is *knowledge*.

Data	Information	Knowledge
54	54 boxes of large cans of baked beans in stock.	Below re-order level – need to re-order baked beans.
£13,347	£1,458,221 total weekly car sales for southern region.	Amount exceeded sales target by over £100,000 for third week running.

Activity 1

Give three examples of your own (not the ones on the right) of data, information and knowledge.

WOLVERHAMPTON COLLEGE

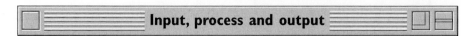
Input, process and output

Information and Communication Technology (ICT) is normally associated with computers. In fact ICT means using any form of modern technology for the collection, storage, processing and sending of information. This can be using computers or other devices such as mobile phones or satellite television.

All ICT devices work on the basis of three stages – input, process and output.

Input means entering **data** into the computer. It is data as it does not have a context at this stage, for example it might be a bank account number, the number from a bar code, a reading from a sensor or a string of letters.

Processing means manipulating this **data** into **information** in a form understandable to the user. This might be by looking up the details of the product whose code has been entered or by performing a calculation on numbers that have been entered.

Output means presenting this information **to the user** or the outside world. It must be in a form the user can understand, so it must have a context. It is information. It could be printed, displayed on a screen or in another form. It might be audible such as an alarm telling the user to evacuate the building.

For example, in a school or college attendance might be collected and input into a computer at each class. At the end of the week all the collected attendance data is processed. This might include summarising,

---

Activity 2 Input – Process – Output

Copy the grid below and fill in possible entries for the missing cells.

Context	Input	Processing	Output
Payroll	Employee name, hours worked	Calculate pay, work out tax etc.	Payslip
Household electricity supply			Bill
Examination Board	Marks for exam papers		
Salesman's visits			Day's list of appointments

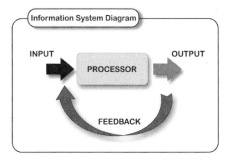

Figure 1.1 Feedback

calculating, storing and sorting the data. At the end of the week tutor group (form) lists are produced showing each student's overall attendance during the week.

ICT also uses **feedback**, which means using the output from a computer to influence the input.

A customer using a self-scanning machine in a supermarket can see the total cost of the goods scanned so far. This **output** may be used to help the customer decide what else they can afford to buy. This is one example of feedback.

> ### EXAMPLE
>
> A pelican crossing, controlled by computer, has a push button device which pedestrians push when they wish to cross the road. When a pedestrian pushes this button it is **input**. A signal is sent to the computer. This is **data** that is input.
>
> The computer calculates how long the red and green light phases should be in each direction, according to instructions in its program. This is **processing** data into a form understood by the user – the changing of the lights.
>
> The lighting of the 'wait' light by the push button, changing of the lights from green to red, turning on the green man and the audible bleep are the **output**. They provide **information** to the users.
>
> A motorist seeing the red light will decide to stop, using the output information to make a decision – this is **knowledge**.
>
> Another pedestrian arriving a few moments later will see that the red man (output) is now lit, preventing her from crossing. She presses the button. This is input affected by the output – **feedback**.

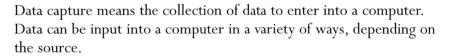

Direct and indirect data capture

Data capture means the collection of data to enter into a computer. Data can be input into a computer in a variety of ways, depending on the source.

Data can be input directly and indirectly. Examples of deriving data directly include:

- a computer reading a bar code in a supermarket
- account details being read directly from the magnetic strip on a credit card
- a computer automatically reading the numbers on the bottom of a cheque
- information from an automatic weather station being down-loaded into a computer.

A VDU operator taking data from a piece of paper and typing it into a database is an example of data being entered indirectly. Indirect entry is more likely to lead to mistakes, due to human error.

EXAMPLE 1

A computer system is used to record the temperature every day at 12 noon.

One way of doing this would be for the operator to go out and read a thermometer at the set time. They then go back to the computer and type in the temperature. This is **indirect data capture**.

Another way of doing this would be to use a temperature sensor and program the computer to read the temperature at the right time. This is **direct data capture**.

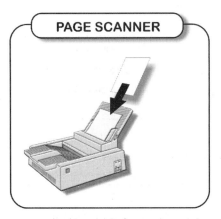

PAGE SCANNER

Figure 1.2 A scanner

EXAMPLE 2

Many businesses use scanners for document image processing. The Consumers Association, publishers of the *Which?* magazine use scanners to read and analyse survey questionnaires.

Surveys in the magazine can generate up to 75,000 responses. Members fill in the questionnaire and return it to the Consumers Association's survey centre in Hertfordshire.

Scanners are used to read the survey results, replacing the old system of reading the questionnaires manually. Using scanners – direct data capture – is faster and more accurate, as well as saving the Consumers Association about £50,000 a year.

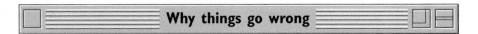
Why things go wrong

If data is entered incorrectly, for whatever reason, (for example, the wrong data has been supplied, it has been mistyped or been given deliberately wrong) the information output will be incorrect. Information is only as accurate as the data that was entered. If the data source is wrong, the information output will be wrong. This is sometimes referred to as **GIGO – Garbage In, Garbage Out**.

Stories abound of things going wrong with computers. A warehouse/ production system at Rootes Group in the 1960s didn't know the difference between feet and inches – some components turned out 12 times bigger than they should have been.

NASA made a similar mistake when trying to send a rocket to Mars. Some measurements were in inches and some in centimetres. Let's just say that the rocket didn't land as expected.

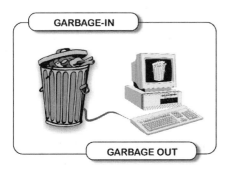

Figure 1.3 GIGO

GIGO

If the wrong data is entered into a computer system, the information that is produced by the system will also be incorrect. It is often the case that the information *appears* to be correct and misleads the person using it. Some examples follow:

'Garbage In'	'Garbage Out'
When entering the marks that a student gained in the different question of an examination, the teacher copies in the marks for the student above him in the list.	The student is awarded a B grade instead of a D in the exam.
A till operator at a supermarket is scanning the barcodes of goods for a customer. The customer has 3 identical tins of cat food. The operator keys in '33' and scans one tin.	The customer receives a bill with a much larger total than expected.
A data entry clerk is recording AS course choices for a new student at a sixth form college. She selects the course from an alphabetic list in a drop down menu. She selects Sport Studies instead of Spanish.	An incorrect timetable is produced for the student.
A householder fills in a card with the current electricity meter reading. Unfortunately, he misreads the first digit as a '5' instead of an '8', thus entering a reading that is less than the last one.	A bill for a negative amount is sent out to the householder.
A consumer orders a new dress by filling in a form for a mail order catalogue. Unfortunately she misreads the number, writing '69' for '96.'	The consumer is sent a pair of shoes.

The last two cases would actually be unlikely to have happened as these errors would have been highlighted by validation checks.

○ Can you list some further examples of GIGO?

Encoding information

Encoding information means putting it into codes. Information is often put into code when stored on computer. For example:

○ gender is usually stored as **M** or **F**

○ banks use branch sortcodes such as **60–18–46**

○ dates of birth such as 6 February 1986, is coded as **06 02 1986**

○ airline baggage handlers use codes for destinations, e.g. **LHR** means London Heathrow, **FRA** means Frankfurt

○ post codes: **SO9** identifies an area of Southampton – **SO9 5NH** is the university. If it is written on an envelope, it can be converted into a series of dots at the sorting office for automatic sorting.

Codes are often easy to remember such as LHR for London Heathrow. However it is not obvious why Birmingham is represented by BHX.

Codes are used because

- the code is usually short and quicker to enter
- the code takes up less storage space on disk
- using a code ensures that the data stored is consistent
- it is easier to check that the code is valid (validation – see Chapter 12).

For example, it is possible to buy a computer program that outputs the full address when you enter a postcode and house number. Entering the postcode:

- is quicker to type in as it is much shorter than the full address
- can immediately be checked to see if the postcode really exists
- avoids spelling the name of street incorrectly.

The data is entered into the computer and stored in code. However if the data item is to become information it must be decoded before it is output. So at a supermarket till, the bar code of the item is entered into the computer as data but information – the name and price of the item – is displayed on the till.

Loss of precision due to coding

Market research questionnaires often ask the subject to tick an age range like this:

a. ☐ Under 18
b. ☐ 18–24
c. ☐ 25–34
d. ☐ 35–44
e. ☐ 45–54
f. ☐ 55–64
g. ☐ 65 or over

The answer is stored as code, e.g. *c* for anyone who is 25 to 34. This coding is easy to use but leads to a loss of precision.

- People who are 25 are bracketed with people who are 34.
- If you had to work out the average age you wouldn't be able to do so.

The fewer the categories, the greater the loss of precision.

Value judgements

A value judgement is when you given a value to something. It is entirely your opinion and may differ completely from someone else's opinion.

Examples of value judgements are if you say something is nice, attractive or ugly.

If someone says, 'I get up at 7 o'clock each morning', this is not a value judgement but a statement of fact. If they say, 'I get up *early* each morning,' this is a value judgement.

A milkman used to getting up at 3.30 wouldn't describe 7 o'clock as *early*. A student used to getting up at 11 o'clock might describe 7 o'clock as *very early*.

Coding value judgements

Businesses are interested in our opinions. Do we like the taste of a product? Is the packaging attractive? Is it too expensive? These are some of the questions a market researcher might ask members of the public before a new product is launched. Information from this research will later be used to determine the price, the image and even whether the product is produced at all. It is likely that there will be a limited choice of responses, such as:

> This new product will cost £1.65. Is it:
> 1. ☐ much too cheap?
> 2. ☐ too cheap?
> 3. ☐ about right?
> 4. ☐ too expensive?
> 5. ☐ much too expensive?

What happens if the customer thinks it is a little bit too expensive? Do they choose 'about right' or 'too expensive'?

Coding value judgements like this leads to a limited number of answers, none of which may be appropriate. The results may not be meaningful because different people will have different opinions on what 'too expensive' and 'much too expensive' mean.

Summary

- data means raw figures
- information means data with a context which gives it a meaning
- knowledge means using information to make decisions
- data can be collected directly or indirectly
- a poor data source will lead to poor information
- information is often encoded when it is stored on computer
- coding information reduces the accuracy and may make it meaningless

Knowledge, information and data questions

1. The expression 'Garbage in, garbage out', or 'GIGO', is often used in connection with information processing systems. Explain, using an example, what is meant by this expression. *(4)*

AQA ICT Module 1 May 2001

2. Explain what is meant by Information and Communication Technology. *(3)*

AQA ICT Module 1 May 2001

3. Information processing is concerned with:

- input
- processing
- output
- feedback

a) Briefly describe these four elements of information processing, using a diagram to illustrate your answer. *(6)*

b) Explain the difference between 'knowledge' and 'information'. *(2)*

AQA ICT Module 1 Specimen Paper

4. With the aid of an example, describe one problem which may occur when coding a value judgement. *(2)*

NEAB 1996 IT01

5. Three components of an Information Processing System are *input*, *processing* and *output*. State what is meant by:

a) input
b) processing
c) output

and give an example of each one. *(6)*

AQA ICT Module 1 January 2001

6. A hotel asked customers to score their service in the hotel according to this system.

1 – Excellent 2 – Good 3 – Average 4 – Bad 5 – Poor

The average mark for 100 customers was 1.8.

Are these statements true?

i) All our customers think our service is good or better.
ii) Our average score is good to excellent.
iii) Our customers think we are consistently good.

Give two reasons why this is not always a reliable way of storing information. *(5)*

7. Encoding value judgements can have the effect of reducing its accuracy or meaning. This becomes evident when the data is retrieved and used. Explain, with the use of two appropriate examples, why this may happen. *(4)*

NEAB Specimen IT01

8. What is meant by the term 'data'? *(1)*

What is meant by the term 'information'? *(1)*

Give an example which clearly shows the difference between 'data' and 'information'. *(2)*

NEAB 1996 Paper 1

9. Many market research firms use questionnaires as a means of gathering raw data for companies about the popularity of their products.

a) Explain why Information Technology is widely used in market research. *(4)*

b) Once the data has been collected, it can be used to give the clients information about their products. Explain the difference between information and data in this context. *(4)*

NEAB 1999 IT01

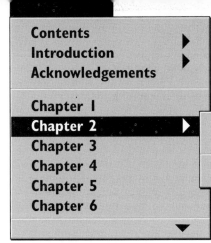

The value and importance of information

Good information

Suppose you see today's weather forecast in a newspaper. You will want the forecast to be:

- **up-to-date** – yesterday's weather forecast will not be much use

- **complete** – a forecast for just the morning may not be sufficient

- **relevant for intended use** – a forecast for Moscow will not be much use for someone in Britain

- **comprehensible** – the forecast must be understood by the reader

- **accurate** – the forecast is not much use if it is inaccurate

Information does not have to be 100 per cent accurate. It has to be accurate enough for the purpose. For example, the exact temperature now may be 16.142356... degrees Celsius. As far as a weather report is concerned, it is accurate enough to say the temperature is 16 degrees.

The source of information is important. The source should be **reliable**. A weather forecast based on information from the meteorological office should be more accurate than that based on an old wives' tale.

The **level of detail** is important. A forecast which takes up a whole page would be too detailed and you may not find the important parts. However a one-sentence forecast may not be in enough detail to be accurate for your area. If you were a sailor setting off on a long single-handed voyage you would require a very detailed forecast.

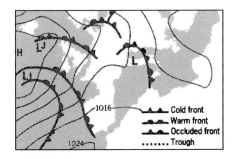

Figure 2.1 The forecast may be accurate and suitable for someone with an understanding of meteorology but is it comprehensible or relevant to the casual reader?

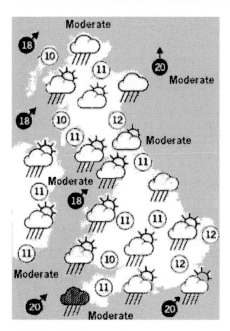

Figure 2.2 This could be better information for the average reader.

The **language** that is used is important. Words like cloudy and sunny are easy for the average reader to understand. Terms like isobars, troughs and occluded fronts may not be.

Today information is a commodity like oil, gold or wheat. It can be very valuable. Of course its value depends on what the information is, how accurate it is, when you get it and what you intend to do with it. Exclusive information to which no other person or organisation has access, is even more valuable.

If businesses are to be successful they need information to make the right decisions. Examples of information businesses might need are:

- sales figures to tell a company how many products to stock
- financial information to plan ahead
- details of competitors' prices to decide on your prices
- names and addresses of customers to contact by direct mail.

Activity 1

For each of the following situations, explain what the consequences could be if information was used that was out of date, inaccurate or incomplete.

1. The purchasing manager of a mail order company is looking at a list of stock levels for his products before placing orders before the Christmas rush.

2. The manager of a holiday company has to pre-book places in hotels for next season. He has the results of a market research survey of holiday preferences from members of the public.

3. A doctor is using the results of a patient's blood test to determine what treatment to give.

4. The board of a manufacturing company has to close down one factory. They have production figures for each factory to help them make their decision.

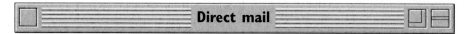

Direct mail

The marketing department of a business is likely to store information on potential customers. Their names and addresses will be stored on a computer file, possibly with other information about them, such as their interests and how much they earn.

The company can then use this file to send out direct-mail letters using *mail-merge*; a program that merges names and addresses into a letter. The mail-merge can be selective; it will select only certain names according to certain criteria. The letter can vary so that people with different interests and different incomes can be told of different products.

Direct-mail enables companies to **target** customers. The idea is that letters go to the people who are likely to be interested in the product and cuts out the people who won't be interested.

Of course direct-mail has increased enormously in recent years and is now said to represent over one sixth of all mail. Some people call it junkmail. Many people do not even open a direct-mail letter – let alone read it! However direct-mail is a very cost effective sales technique.

Case Study 1

Porsche Cars, Reading

Porsche Cars Great Britain (PCGB) imports all Porsches into Britain and owns five Porsche dealerships. Sales are small compared with volume manufacturers such as Ford or Vauxhall.

PCGB know that the people most likely to buy one of their cars are people who have bought one in the past. These people obviously have an interest in sports cars and presumably have the necessary income to purchase one.

This means that PCGB spend 80 per cent of their marketing budget on direct-mail targeting previous customers. They have built up a large database of 22,000 current Porsche owners, over half of all Porsche owners in the country. These owners receive a copy of *Marque*, PCGB's magazine. This is an important part of their marketing strategy. Industry sources suggest that sixty per cent of customers will buy from you again simply because you keep in touch.

Keeping a database like this is quite legal but PCGB must comply with the Data Protection Act (see Chapter 9).

- Give three reasons why PCGB use this method of promoting their cars rather than advertise on television as volume car manufacturers do.

- Give three further products, other than cars, that might best be marketed in this way.

Companies collect information on their customers. This might be through past sales like PCGB. Supermarkets such as Tesco and Sainsbury's each have information on around 12 million customers through their loyalty cards.

However to receive a card a customer must fill in a form giving personal details. These details are then stored on computer. These cards give discounts on products.

Whenever the customer buys a product, this information too is stored on computer. This enables the supermarket to build up a picture of customers' shopping patterns.

For example, they can see what time you go shopping, whether you buy meat or vegetarian products, whether you buy baby products. From these they can deduce a lot about each customer. As one customer said, 'They know more about my shopping habits than I do.'

Whenever goods are purchased by mail order or on-line via the Internet data about the customer is gathered; whenever a form requesting a 'special offer' for goods or information is cut out from a newspaper or magazine, the consumer's details are likely to be stored. Organisations have to ask customers if their data can be passed on to others. This is often done by using a tick box, often placed in an obscure part of the form in very small text that gives a message such as:

☐ Tick here if you do not wish to receive mail from other companies

Information gathered in this way can be used to target specific customers through direct mail.

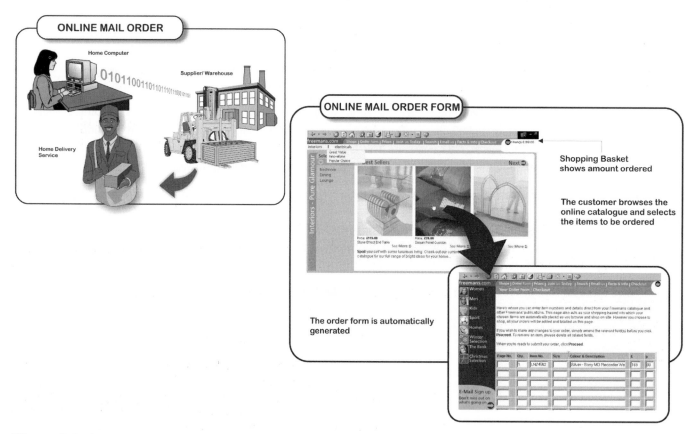

Figure 2.3 On-line mail order

A lot of information is freely available to the public. The electoral register is compiled by local councils and lists the names and addresses of people entitled to vote in elections. A section might look like this:

Paddock View, Pillsbury, P85 3RH

1 Paddock View	Albert Mitchell
1 Paddock View	Doris Mitchell
2 Paddock View	Sally Bryant
3 Paddock View	Graham Williams
3 Paddock View	Sally Williams
3 Paddock View	Daniel Williams
3 Paddock View	Adam Williams (*10 August*)
4 Paddock View	William Hunter
4 Paddock View	Shirley Hunter

There are a number of things we can deduce from this information which businesses might find valuable.

1. All nine people live in Pillsbury. A local restaurant might see them as potential customers.

2. Graham Williams has a large family. He might be interested in a large car or a home extension

3. A date appears after Adam Williams' name. He will be old enough to vote on that day and so is approaching 18. He might be a potential customer for local night-club.

4. The names of the residents of 1 Paddock View suggest they are likely to be elderly. They might not be interested in a local night-club but could be interested in, say, Saga holidays.

If we also know that Paddock View consists of owner-occupied houses built in the 1990s.

5. All residents may be interested in home improvements such as double-glazing as they own their own homes.

Remember: We can deduce all this from just a list of names and addresses.

Activity 2

Can you deduce anything else from the list of names and addresses?

Overheads: the cost of obtaining information

Information is valuable and there are costs involved in collecting it. Even when information is available for free, for example the electoral register, there are labour costs in entering data and converting it into the right format.

The electoral register on paper is not very convenient. The storage space needed to store every voter in the United Kingdom would be enormous. Searching for individual voters would be nearly impossible. If the data were to be stored on computer, it would have to be typed in.

Companies like Millennium Data Ltd (http://www.marketinglists.com/eroll.htm) resell the electoral roll in disc format. They say this electronic format is useful for

- direct marketing campaigns

- political campaigns

- data analysis, capture and validation

- software development.

Buying the electronic version of an electoral register from this company will be more expensive than looking at the free, paper based registers but it will be more convenient.

The more information that is required and the more detailed the information, the more it will cost to obtain. For example, if the data is collected through a survey, the cost will depend on the size of the sample and the number of questions asked.

Postcode information is very valuable to companies. Companies like AFD software (http://www.afd.co.uk/) sell software that finds out address information from postcode data. Sales staff can find addresses quickly and accurately by entering the postcode.

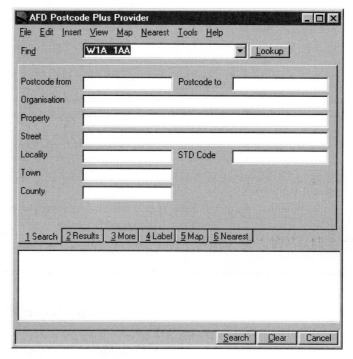

Figure 2.4 Postcode Plus Provider allows a user to enter a postcode and display the address

AFD say that the software enables users to

- look up quickly the full address from any part of the address

- save over 80 per cent of the keystrokes when entering addresses

- save hours of wasted time

- check correct postal addresses

- eliminate errors in your address lists

- print labels or envelopes for any address

- print in postcode order to take advantage of discounts for sorted mail.

You can now find addresses from postcodes for free on the Internet at http://www.royalmail.co.uk/paf/

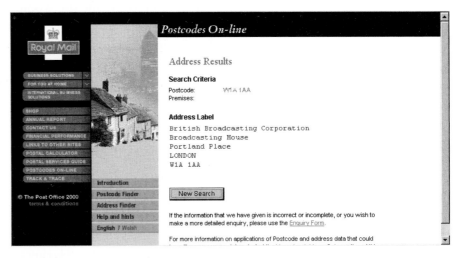

Figure 2.5 Postcode

However buying dedicated software is likely to provide faster access to the information and more features.

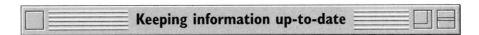
Keeping information up-to-date

Information has to be up-to-date to be useful. Last week's weather forecast is not much use to us now. New houses are being built in new streets all the time. The postcode database needs to be kept up-to-date. Keeping the information up-to-date will also affect the costs. It will involve

- collecting the data more often, so increasing the costs

- entering the new data

- deleting any old data.

Information may be out-of-date by the time it has been processed. Up-to-date data may require faster processing, more sophisticated hardware and faster communications. This will also increase the costs.

Lost information

The value of information is best shown by the cost of losing it.

Just imagine what would happen if a supermarket lost the file storing bar-codes, product names, prices and sales figures. Not only would the check-outs not function but the stock re-ordering system would also not work either.

The cost of losing the file would not just be the wages involved in collecting the data and entering it into the computer again but also the lost business while the information was unavailable.

A credit card company calculated that for every hour its computers were out of action and the information to authorise sales was not available, the company would lose $2.6 million.

Disaster recovery experts say that a company experiencing a computer breakdown lasting more than ten days, will never fully recover financially. Half of these companies will go out of business altogether. No wonder companies regard information as valuable.

A study in the USA estimated that six per cent of PCs suffer from data loss in any given year. Each incident cost an average of $2,500 to rectify, including the cost of retrieving and recovering the missing information and lost productivity. This would mean that American businesses spend $12 billion a year recovering data.

Sometimes loss of data is not just inconvenient but very embarrassing…

Case Study 2

Power surge wipes patients' names off hospital transport list
Burton Mail 17 May 1999

An urgent appeal to disabled and elderly people was going out today after computer chaos wiped up to a thousand patients off a hospital transport list.

Fears have been raised that patients may miss out on their ambulance and taxi journeys to hospitals, day centres and clinics because their names have been scrubbed from a computer.

Staffordshire ambulance bosses blamed a sudden 'power surge' which blacked out its electricity supply on Friday.

They say this has removed between 800 and 1,000 cases from their non-emergency passenger transport lists (PTS).

- What could the consequences be of this data loss?
- Do you think that the Ambulance Service should have done anything to prevent the loss of data? If so what kinds of steps could they have taken?

Activity 3

Use the Internet to research other examples of organisations losing their data. A good starting point would be a newspaper site that allows you to search for keywords in its stories. The following site allows you to search magazine stories and could also be a useful starting point:
http://www.computingnet.co.uk

Obtaining information illegally

Information can be so valuable that reputable companies may take illegal steps to get it.

In 1991 Richard Branson's Virgin Atlantic Airways accused British Airways (BA) of using such illegal 'dirty tricks' to gain access to information on Virgin's computer files, particularly the names and addresses of Virgin's business customers. Branson alleged BA then contacted these customers offering big discounts and other incentives to switch from flying Virgin to flying BA.

In March 1992 Branson sued BA for libel over suggestions that he had invented the claims. In January 1993 BA apologised unreservedly to Branson and paid Virgin Atlantic over £1.5 million in costs and damages in an out-of-court settlement.

'Industrial espionage' like this to obtain information from a competitor is not unusual. Accurate information would be very useful to a rival company. The value of commercial information depends on its accuracy, the potential user and the intended use.

Control of information

If information is valuable, as with any commodity, it is important that is kept securely and access to it is controlled.

As we have seen, theft of computer data is a real possibility. Every major company in the UK has been the target of some attempt to steal data stored on its computers.

In many organisations different users will need different access privileges. For example a library computer stores details of books, members of the library and books borrowed.

- Members can use the system to search for particular books but cannot see the personal details of members.

- Some library staff can check up on who has borrowed which book if it is overdue. They can read but not change personal information on members.

- Other library staff can enter the details of new members joining the library, edit old details, e.g. when they move house and delete details of members who have left or died. They can edit personal information on members.

The Data Protection Act (see page 96) says that personal data stored on computer must be kept secret. It is likely that computer users will want to go further than this and control access to all sensitive data, not just that about people.

Summary

- good information should be accurate, up-to-date, comprehensible, relevant and complete

- information can have a monetary value

- the value depends on its potential use and its intended use

- to be valuable information must be correct and up-to-date

- information in electronic format may be more useful and so more valuable than hard copy

- ensuring information is up-to-date can be time consuming and costly

Value and importance of information questions

1. State **three** factors that affect the value and importance of information. Give an example that shows clearly how each factor affects the information's value. *(6)*

   AQA ICT Module 1 January 2001

2. Travelling sales representatives working in the UK can make extensive use of company credit cards to pay for goods and services. A company credit card is one that is issued by a company to its representative. All charges and information relating to each transaction are sent directly to the company.

   a) List four items of data which are captured each time the card is used. *(4)*

   b) Other than payment information, suggest one other potential use for information which can be derived from this data. *(2)*

   NEAB 1997 IT01

3. Low quality information can be misleading, distorted or incomprehensible. This type of information is of little value to the decision maker. The output of good quality information is costly and dependent upon many factors.

   a) Identify three factors which affect the quality of information. *(3)*

   b) State two factors which affect the cost of providing good quality information. *(2)*

   AQA ICT Module 1 Specimen Paper

4. State two factors that affect the value of information and give an example of each one. *(4)*

   NEAB 1999 IT01

5. A supermarket stock control computer system up-dates its stock levels every evening based on that day's sales. List two possible consequences of the supermarket using out-of-date data. *(2)*

6. A shop has between 100 and 500 customers who have accounts. These customers have papers delivered and magazines ordered regularly. Their accounts are payable weekly.

   a) By giving examples from the customers' accounts, explain the difference between data and information. *(2)*

   b) Describe two factors that may make the customers' accounts **data inaccurate** and explain how this affects the **quality of the information** obtainable. *(4)*

   OCR ICT Specimen

7. A travel firm arranging package holidays in Spain for the 2001 season uses data obtained from a survey of their customers' favourite holiday resorts in 1997. Explain

   a) why the data from 1997 might not be suitable for use to predict the requirements for 2001

   b) what the effect on the company might be if it used the 1997 data. *(4)*

   AQA ICT Module 1 May 2001

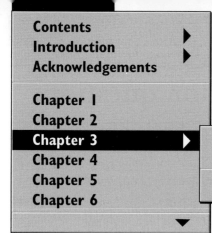
Capabilities and limitations of ICT

Information processing machines

Computers are machines that process data and turn it into information. What is it about computers that has meant that they play such a large part in twenty-first century life?

- **They are fast**. Computers process data very quickly, certainly much, much faster than a human could do it. This means they can also process large volumes of data. Not only that but they keep getting faster with the processor speed doubling about every 18 months. As computers get faster there are more and more applications for which they can be used. Accurate, detailed weather forecasts could not be produced without very fast computers that run simulation programs that manipulate data on many different weather factors. To be of any use, a weather forecasting program has to run faster than real-time – it is no good getting a forecast for Monday morning's weather on Tuesday, or even on Monday afternoon!

- **They can search very quickly**. Computers can search through these large volumes of data very quickly. For example, files can be located based on the name of the file, the date of creation or particular text stored within the file. Consider a computer system used by the police that stores details of the fingerprints of hundreds of thousands of people in digital form. When a fingerprint is detected at the scene of a crime a search needs to be made to see if the print matches any that are held. Without a computer this task could not be undertaken for all the thousands of stored prints within a realistic time.

- **They have vast storage capacity**. Vast amounts of data be stored either in the computer's memory or in a backing store medium such as the hard disk or CD-ROM. For example, whole encyclopaedias or a year's newspaper articles can easily be stored on one CD-ROM. Today even a modestly priced home microcomputer has sufficient storage capacity to enable a user to run a program such as an action game that has very realistic animated graphics. Such graphical images require large amounts of storage space.

- **They are very accurate**. Although there is the possibility of hardware failure or software bugs, computers are much more accurate in processing data than humans because they can perform repetitive tasks without becoming bored or tired. If a piece of software such as a spreadsheet is set up to perform a task such as adding up a list of numbers, it will come up with the same result every time the same figures are entered. A human, on the other hand, is quite likely to get the result wrong sometimes, particularly if they have been working for a long time and are tired.

- **They perform repetitive tasks well**. Many computer applications involve repetitive tasks such as printing bank statements or calculating wages for employees. Paying wages for 5,000 people is no harder than paying wages for five people when using a computer.

- **They are automatic**. Computer systems work automatically and need little human supervision. They can work 24 hours a day, 7 days a week and can be programmed to perform certain tasks without an operator being present. For example, many organisations will arrange for file backup to be carried out automatically at night.

- **They can combine data**. Data from different sources can be combined to provide high quality information in a variety of output formats. In a school or college a student attendance system collects data on the presence of each student in each of his classes. The data from all classes can be combined and sorted to produce a weekly summary for each student of his attendance in all classes.

- **They can link to other computer systems**. Computers can link to other computer systems and other electronic devices, almost anywhere in the world. This has also increased the number of applications for which computers can be used. The Internet allows users to communicate with other users world wide as well as access huge amount of data through the **World Wide Web**. A holiday company will store details of all available holidays together with bookings that have been made on a central computer. This will be available to travel agents in many different locations who can link their computer to the central data store. This enables a travel agent both to have access to up-to-date information on holiday availability, and to make immediate bookings while their client is with them.

All these characteristics mean that computers can be used in many situations today and provide information of a high quality. Many tasks currently carried out were impossible before the development of computers. Accurate weather forecasting, managing large supermarkets and processing financial transactions are all tasks that could not have been performed without computers.

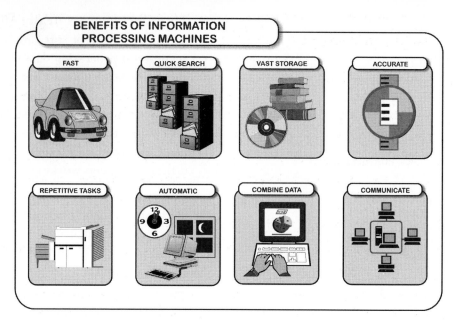

Figure 3.1 Characteristics of information processing machines

Each of the following six case studies describes a particular computer application. Read each carefully, then answer the questions.

Case Study 1

The Police National Computer

The Police National Computer (PNC) stores data needed by the police for its operations. This includes details of:

- criminals
- vehicles and their owners
- stolen property.

The PNC was designed to have a response time of a fraction of a second. It is capable of storing 5 million criminal records and 50 million vehicle records. The PNC handles around 200,000 accesses per day from UK police forces, updating and searching files. That's over 65 million in a year.

The PNC can be accessed from police stations up and down the country, 24 hours a day. The police are now introducing mobile equipment, which enables police officers to access information when away from the police station.

- Describe the advantages to the police of the PNC over a non-computerised system.

Case Study 2

Public Lending Library

A lending library in Wintown has a very large stock of books, CDs, audio tapes and videos that are loaned to members for a fixed time period. The computer system used to manage the flow of data in the library is linked in via a wide area network to the central county computer.

The computer records details of all loaned and returned items. Whenever a loan is made the system checks that it will not exceed the member's allowable loan limit; also the member will be reminded if he has any overdue items.

An electronic catalogue of all books held by the library is maintained for the use of both staff and members. The central county computer can be accessed to extend the search for a book countywide.

Once a week a mail merge program is run that produces letters to be sent to all those members who have items that are over a month overdue.

Every six months a summary is produced for the senior librarian that provides statistics on borrowing patterns. A list is also produced that contains the details of all items that have not been borrowed during the last period.

- What is a mail merge?

- Refer to the list of capabilities given at the start of the chapter. Explain which apply to the Wintown library system.

Case Study 3

Gas Billing System

A regional gas company delivers gas to over half a million subscribers. Details of all customers are held on the computer and every three months bills are printed out and sent to the appropriate householders. Each bill is worked out from a reading taken from an individual property.

Details of any payment made are used to update a customer's record. If payment is not made within a defined period the customer is sent a further bill as a reminder.

Each meter reader has a hand held device that he uses to record the number read from the meter. The device already contains details of the customer and house. At the end of the day the meter reader attaches his device to a connecter that is installed at his home. This is then connected via a telephone line to the company's wide area network. Details of the readings he has made are transferred to the central computer and details of the households that he is to visit the next day are downloaded to his device.

- What information, apart from the recent meter reading, will be needed to produce a customer's bill?

- Refer to the list of capabilities given at the start of the chapter. Explain which apply to the gas billing system.

Case Study 4

Sunnyside Hospital Patient Administration

Sunnyside hospital maintains linked records of all outpatient appointments and inpatient stays. Details are linked by the use of a unique patient number. There are records of over 200,000 patients in the system which has been in use for several years.

A number of different people add data relating to the patient to the system:

- a clerk records details of appointments and visits to outpatients

- the ward clerk enters details of a patient's stay in a ward

- the radiographer stores details of X-rays

- laboratory staff enter results of blood and other pathology tests carried out

- ward nurses and doctors enter details of medications issued and treatments undertaken.

Doctors can refer to all the information relating to a patient under their care.

- State the advantages of using the computer based system described over a manual one.

- Why is each patient allocated a unique patient number?

- Refer to the list of capabilities given at the start of the chapter. Explain which apply to the patient administrative system.

Case Study 5

Aid Agency

An aid agency provides help to people in need around the world after a disaster such as a flood, earthquake or famine has taken place.

Local field stations are set up and whenever possible these are equipped with a computer. This allows for fast communication links so that e-mails can be sent giving details of the latest situation and needs,

The agency maintains a database of skilled volunteers such as doctors, engineers or builders who woud be prepared to travel to help out if an emergency were to arise.

A further data base containing details of potential donors is maintained and when an emergency does occur a mail merge is used to circulate them asking for donations.

- Refer to the list of capabilities given at the start of the chapter. Explain which apply to the aid agency systems.

Case Study 6

Airline Bookings

Every airline maintains a centralised booking system for its flights; travel agents all around the world can access the system to find out about seat availability as well as make a booking. Bookings can be made several months in advance. For many large airlines there will be thousands of flights available for booking at any one time.

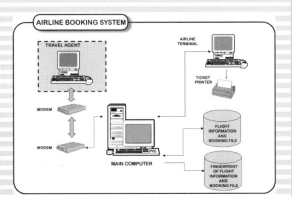

Whenever a booking is made a number of details relating to the passengers is recorded, the total cost including airport fees and other charges is calculated and an invoice printed. Bookings can be made right up to the last minute.

Many systems will allow the travel agent to print the ticket for the customer at the time of booking.

Details of every flight are also recorded. These will include details of the date and time of the flight, the departure and destination airports, the crew and the plane to be used.

When the passenger arrives at the airport he checks in and is allocated a seat number. His luggage is also booked in and has a label attached that holds details of airport destination along with an identification code. All the details are recorded on the computer.

A number of lists need to be produced which include:

- a full passenger list for check-in clerks
- a list showing special diet requirements for the caterers
- a list of passengers needing special service together with their seat numbers.

Flight availability can be checked and bookings made over the Internet.

- To what use could the baggage code be put?
- Refer to the list of capabilities given at the start of the chapter. Explain which apply to the airline booking system.
- What problems could arise from having a booking system that is dependant upon computers?

Response speed

The speed of response is very important in many modern computer systems. A police officer might want to know quickly if a particular car is stolen. When withdrawing money from a cash machine a customer does not want to wait a long time for the computer to process the data and check whether enough money is available in their account.

Fast response times are important in systems which involve feedback. For example, a stock exchange system allows shares to be bought and sold electronically. Share prices go up and down depending on how many have been bought or sold. Share prices are adjusted automatically.

This is feedback – the output (details of shares bought and sold) affects the input (share prices). Fast response times are essential for this to work successfully.

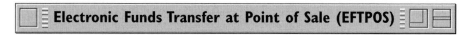

Electronic Funds Transfer at Point of Sale (EFTPOS)

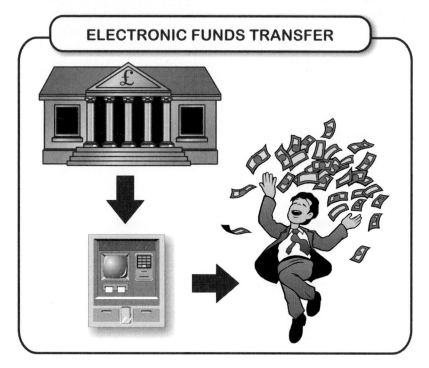

Figure 3.4 Electronic Funds Transfer

Payment for goods can be made in a number of ways:

- by cash
- by cheque
- by credit or debit card.

The use of computers has allowed for the development in the payment of goods by means other than cash. Many payments are made electronically – in shops and supermarkets, ticket offices and restaurants.

Banks throughout the world have large mainframe computers that process customer accounts. Software that enables Electronic Transfer of Funds (EFT) to take place has been written especially for these computers. The international nature of banking, with banks in every time zone, means that EFT systems run 24 hours a day.

When a person wishes to make a payment using EFT, his card is first 'swiped' through a reader that reads the details from the magnetic strip. The reader is linked via a wide area network to a central bank computer that checks the details of the proposed transaction against the state of the individual's bank account. If there are sufficient funds in the account to cover the payment an authorisation code is sent back. The response needs to be returned within a few seconds otherwise the customer will become impatient and a long queue may build up at the point of sale.

If the response authorises the payment the sale is then completed. This process is an example of feedback: the salesperson is able to carryout the sale in the knowledge that the payment is authorised by the bank.

The software written for EFS is capable of handling large numbers of transactions per second.

Limitations of information and communication technology

Although modern-day computers have phenomenal capabilities, there are limitations in the use of ICT systems. The nature of these limitations will vary over time.

At any time, there are tasks that cannot be carried out adequately using the hardware that is currently available. Until processor speeds and storage capacity reached certain levels the display and storage of extensive graphics and sound was not possible. Until a bar code scanner was developed automatic data capture at a point of sale terminal could not be undertaken.

Even when appropriate hardware is available, failure is not uncommon. The result of such failure, if no suitable alternative plans have been made, can be severe. Disk failure can lead to the loss of data if backup copies have not been kept.

Faster and more reliable processors are being developed all the time.

Computers will only process the data that is entered into a system. If the wrong data is entered into the computer, the wrong information will come out. The entry of incorrect data is usually a human error, although the incorrect results are often blamed on the computer! The need for appropriate data control mechanisms that detect such human error are vital if a computer system is to produce reliable and accurate information. However, it is difficult to prevent or pick up all human errors.

Software bugs can make a system fail or behave in an unpredictable manner. This can happen because the software has been rushed onto the market without being fully tested. This is often done for commercial reasons, perhaps to beat a competitor.

Poorly designed systems will not achieve what they set out to do. A system can be poorly designed as a result of inadequate investigation. Perhaps some possible data inputs have not been considered or the results have been calculated in the wrong way. Remember the computer is only a dumb tool and will do what it is programmed to do. If the programming is wrong, the output will be wrong.

Processes that are easy for humans are hardest for computers. A human finds it easy to recognise images as well as understand both spoken and written language. These are all tasks that a computer finds very hard.

Specialists working in the field of Artificial Intelligence (AI) have developed software that can carry out some of these tasks. There are programs that will read handwriting for example, but no computer can yet perform as well in this area as a typical human.

It takes a long time to develop a new system. Sometimes a system is out of date before it is even used.

There have been huge developments in communications technology over the last few years. However communications can still cause a bottleneck in a system.

The speed of transfer across a data communication link is often inadequate to allow for the volume of data that needs to be transferred. However technological developments continue; broadband technology, which means increased Internet bandwidth and thus faster data transfer, offers the hope of faster connections becoming commonplace.

The Internet

The Internet has been re-christened by some as the *Word Wide Wait*. You will probably have experienced:

- waiting for web pages to load through a phone line and a modem.

- some pages and/or images not loading at all

- the computer crashing or the connection being lost while you are using the 'net'

- having to wait even longer in the afternoon or evening because America is '*on-line*'

- not being able to connect to your Internet service provider because the lines are 'busy'

Although technically feasible and already used by many, the growth of e-commerce has not been as fast as some people predicted. This is partly due to a lack of consumer trust in the security issues of transferring credit card details via the public telephones of the Internet.

Activity 1

Visit a newspaper or magazine site on the Internet and search for 'computer errors'. Find some examples. For each example determine the cause of the error.

Case Study 7

Daily Mail, June 26 2001

The great giveaway

How money-off blunder at Sainsbury's handed couple free groceries worth £900

By Sean Poulter Consumer Affairs Correspondent

When supermarkets make mistakes over pricing, it seldom seems to work in the customer's favour.

But try telling that to Rebecca Sheldrake and her husband Pete.

Due to a computer error linked to a hair care product offer at Sainsbury's, the young couple found themselves walking out of their local store with free groceries worth hundreds of pounds. Unable to believe their luck, they did a supermarket sweep of every Sainsbury's in their area and saved around £900.

The error surrounded a promotion for the John Frieda hair care range Frizz-Ease. Customers were supposed to get a free bottle of shampoo worth £3.50 when they bought one other product from the range. But Mrs Sheldrake began racking up the savings when she realised that every time two sachets of the £1.25 conditioner were put though the computerised checkout, one came free – and £3.50 was being knocked off her shopping bill. Obviously less than pleased by their bad hair day, Sainsbury's staff challenged Mr Sheldrake, 31, a fork-lift engineer, as he loaded up in the car park of one store in Hedge End, Hampshire.

'They weren't very polite to him,' said his 25-year-old wife. 'There must have been 20 members of staff and all the other shoppers staring. He was extremely embarrassed.' Managers told him he was not entitled to the savings as there was a limit of six sachets per customer, despite the fact that there was no such stipulation on the promotional material.

But once the couple took their case to trading standards officers, Sainsbury's backed down – and offered the couple a further £350 to make up for the embarrassment of being forced to return the shopping. The Sheldrakes now find their cupboards stuffed with goods while their daughter Paige is looking forward to a few extra presents as her third birthday approaches.

The computer glitch came to light when the couple were shopping at their local Alder Hills store in Bournemouth.

'I was originally going to buy two sachets of conditioner, but in the end I took every one in the store,' said Mrs Sheldrake, a housewife. 'I have got wardrobes full of the stuff now.

'To be honest it is a struggle every month to pay the shopping bills. So we concentrated on things that are expensive but can be stored, such as nappies and washing powder. We did get the odd extra cake or bar of chocolate as a treat, but there was no champagne or anything like that.'

The couple decided to treat Paige to a dressing-up outfit for her favourite TV character, Bob the Builder. 'The toys can be very expensive and I would not normally have bought them, but this gave us a chance to get something she wanted,' said Mrs Sheldrake.

Over three weeks, the family made 12 visits to six different Sainsbury's stores in Dorset and Hampshire.

The offer on Frizz-Ease has come to an end. A Sainsbury spokesman said: 'To ensure this doesn't happen again we have revised our in-store promotions.'

- Was the cause of the error due to the computer or the human?

- Explain in your own words how the Sheldrakes managed to acquire so many free goods.

Summary

Computers

- are fast
- can store and process vast amounts of data
- can search very quickly
- are very accurate
- perform repetitive tasks well
- are automatic
- can work 24 hours a day, 7 days a week
- can be programmed to perform certain tasks at night without an operator being present
- handle large volumes of data

However, there are limitations to the use of ICT.

Factors include:

- hardware
- software
- communications
- inappropriate design
- poor data control mechanisms

Capabilities and limitations of ICT questions

1. A national distribution company advertises its products by sending personalised letters to thousands of people across the country each year. This type of letter is often known as 'junk mail'. The distribution company purchases the list of names and addresses from an agency.

   a) Describe two ways in which the use of information technology has increased the use of 'junk mail'. *(4)*

   b) The company wishes to target letters to people who are likely to buy its products. How might this be done? *(2)*

   NEAB 1996 Paper 1

2. When incorrect bills are sent to customers, an organisation often gives the reason as 'The computer got it wrong'. Using an example, give a more likely explanation. *(4)*

   AQA ICT Module 1 January 2001

3. Describe two advantages to a company of sending out direct mail letters to customers rather than sending a standard letter beginning 'Dear Sir'. *(2)*

4. Most companies now use computers to calculate their employees' wages and pay them straight into their bank account. Suggest four reasons why they no longer calculate wages manually. *(4)*

The social impact of ICT

Computers have already had a great effect on society, for example in the fields of employment, leisure and communications. This trend is likely to increase as processors get faster, computer memory gets bigger and their price is comparatively cheaper. In this chapter we look at some of the ways IT has affected the way we live and work.

IT and manufacturing

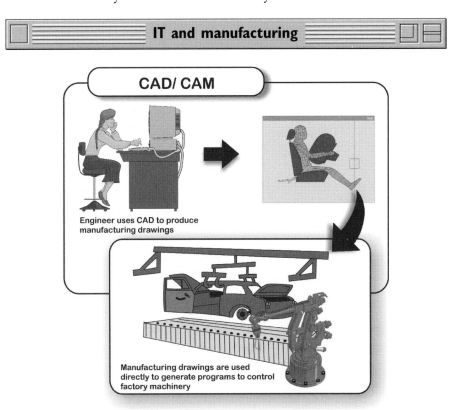

Figure 4.1 CAD/CAM

Many industries now use Computer-Aided Design/Computer-Aided Manufacturing **(CAD/CAM)** where the output from a CAD process is input to control a manufacturing process.

Benetton, the fashion company, uses a CAD system to produce a template for items such as a pair of jeans in a range of sizes. The CAD system automatically calculates the best way to lay the templates on the fabric so as to minimise the wastage of materials used and cuts the time taken to produce these templates from 24 to two hours.

Benetton links its CAD system to its CAM system. This enables designs created on the CAD system to be transferred directly to computer-controlled knitting machines, so increasing the flexibility of the systems still further.

The quality of computer-manufactured products is higher and more consistent, leading to greater reliability and increased productivity. The use of CAM should optimise the use of raw materials, so cutting down on wastage and consequently cutting costs. Working conditions are often cleaner. More automation offers the workforce the prospect of shorter working hours and more leisure time, but it always doesn't work out as predicted. There may be further reductions in the workforce.

Products can also be manufactured with fewer staff. A brewery canning line needed 15 people to operate it before computerisation. Now it needs only two, yet more beer is canned and there are fewer breakdowns. Obviously, some jobs have been lost. Workers have been made redundant or redeployed to other jobs. Some skilled workers may have seen their skills replaced by automatic machines.

Just In Time (JIT) processing is a method by which supplies for processing, such as raw materials, only arrive exactly when they are needed. This reduces the need for expensive storage space and also reduces the amount of money that is tied up in stock. JIT depends upon the existence of an efficient computerised automatic ordering system that only orders parts to arrive exactly when they are needed.

Computers have created jobs in the manufacturing of hardware, and also in sales, repair and maintenance, technical support and consultancy sectors. New products are now being manufactured thanks to computers, such as CDs, videos, microwaves and satellite dishes.

IT and commerce

Business today is unrecognisable compared with business in the days before computers. The electronic office is an obvious example of the effect of ICT on business. A modern office is likely to have a computer on every desk. The work carried out in offices is generally the receipt, processing, storage and despatch of information. The computer can do all these things more efficiently than traditional methods. The use of word processing and desktop publishing packages has meant that even very small businesses can produce material in a very professional manner and so enhance their image with both existing and potential customers.

The arrival of fax, e-mail and satellite links has completely changed how businesses and some individuals communicate. Businesses now advertise their fax numbers and e-mail addresses prominently, which makes it easier for customers to keep in touch. It would seem logical that this would affect the level of traditional communications, particularly letters sent by post.

Products can be ordered on-line using the Internet. Electronic Data Interchange **(EDI)** is a system of sending orders, paying invoices and sending information electronically (see Chapter 5 for more details). Money can be transferred electronically to pay for goods and services. These are all examples of *e-commerce* (electronic commerce), which is revolutionising how businesses order and sell products.

Shopping via the Internet is taking an increasing share of the market. It is not just hi-tech and multinational companies that use e-commerce to sell goods. A family-owned butchers from Yorkshire, Jack Scaife, started to use an Internet site (**www.jackscaife.co.uk**) to sell its bacon, sending deliveries all over the world. Soon e-commerce was bringing in £200,000 worth of sales from a site that cost £250 a year to maintain.

Computerised stock control has enabled supermarkets to reduce costs by reducing stock levels and stock even more products than before. Computers in the shops and warehouses are linked as a wide area network. When a store's stock for an item falls below a pre-set level the store computer automatically sends an electronic message to the warehouse to initiate the transfer of new stocks to the specific supermarket. The warehouse computer will produce a 'picking list', a document used by warehouse staff to select items to be dispatched. In many warehouses the 'picking' is done by a computer-controlled robot.

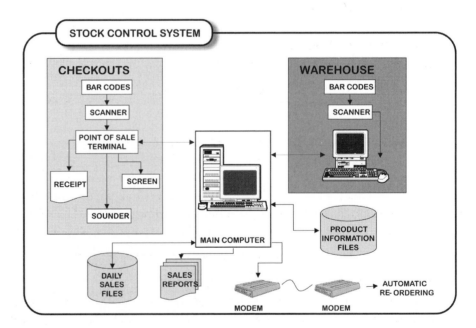

Figure 4.2 Stock Control System

Banking

The number of transactions carried out by banks has grown so rapidly that they could not operate without computers now. Banks transfer money electronically. Most workers are paid directly into their bank account by computer, while many regular payments such as mortgages, utilities and insurance premiums are transferred electronically from an individual's bank account to the company's bank account as a direct debit or a standing order.

Cash is not as important as it used to be. Most individuals do not need to carry as much cash as before thanks to credit and debit cards. If a person does run out of cash, they can visit an automated teller machine (ATM) at any time. Home banking means that users can check their accounts and pay bills from home.

ICT has also affected our shopping habits in other ways. The widespread use of credit cards mean it is possible to shop and pay for goods without leaving home. Cable and digital TV shopping channels and the Internet have provided new ways of finding out what to buy instead of the traditional catalogues.

Case Study 1

The Cashless Society

There have been many developments in information technology that are leading to a society without cash. These include:

- credit cards where computers store financial details
- cheques which are processed by computers using MICR
- direct debit used to pay regular bills is generated by computer
- wages and salaries are paid by electronic transfer and not in cash
- phone-cards can be used for telephone calls
- smart cards containing a microchip can be used for automatic debit and credit (see below)
- electronic fund transfer connects the shop with the banks' computers.

In 1995 a company called Mondex launched an experiment in Swindon hailed as the start of the cashless society. Local people were issued with a Mondex card, described as an 'electronic purse' that can be charged with cash and then used to pay for goods and services, for a monthly fee of £1.50. After three months, just 4 per cent of householders had the card and a total of only £250,000 had been spent using Mondex.

In 1996 a rival company, Visa International, one of the world's largest credit card providers, started a similar experiment in Leeds. Visa Cash, a chip-based plastic card, allowed users to make everyday purchases of small items such as a newspaper or a pint of milk, without having to scrape around for the right change. Holders of the Visa Cash card can 'load' the card's electronic chip from their bank account, up to a limit of £100, at any of 3000 specially designed automated telling machines.

- What do you think are the advantages and disadvantages of a cashless society?

Computers are used in the administration of hospitals and doctors' surgeries, storing patients' records. Pharmacists keep records of customers and their prescriptions. When a patient's records are needed by a health professional they will always be available, unlike paper-based records that can only be viewed in one place at a time. The use of electronic records can lead to concerns about the consequences of storing inaccurate data and the increased threats to the security of the data.

Some hospitals are now experimenting with storing medical records on smart cards kept by the patient and taken with them every time they visit a doctor, dentist, pharmacist or hospital. The smart card can store a complete medical history and can be updated at the end of each visit.

IT also helps in the diagnosis and monitoring of patients' illnesses. Expert systems can be set up to help in diagnosis by asking questions about symptoms and using the answers to draw conclusions. Computer controlled ultra sound scanners enable doctors to screen patients very accurately. X-ray film is being replaced by on-screen digital pictures. Computers can be used for continuous monitoring of patients' bodily functions such as blood pressure, pulse and respiration rates. Such systems provide instant feedback of information and can free up nurses to carry out other duties.

Case Study 2

Use of Robots in Surgery

Robots are now being used in operations undertaken for some human conditions where surgery is needed at a smaller scale than conventional surgery permits. This allows some operations to be carried out with much greater accuracy and success than traditional methods can achieve. Current systems consist of two main components: a viewing and control console and a surgical arm unit.

In using the system for an abdominal operation, tiny incisions are made in the patient's abdomen and three stainless steel rods are inserted. Each rod is held in place by a robotic arm. One of the rods is equipped with a camera, while the other two are fitted with surgical instruments. The doctor's hands do not directly touch these instruments, but sits at the control console and studies the 3-D images sent by the camera inside the patient to be displayed on a screen.

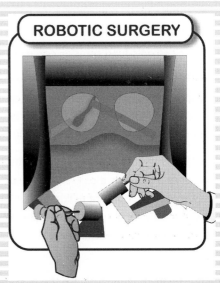

ROBOTIC SURGERY

The surgeon can manipulate the surgical instruments by using controls that are similar to joysticks. The surgeon uses voice-activated software to position the camera.

IT in the home

Computers can be used in the home to control and protect the environment by controlling heating, ventilation, hot water and security devices such as alarms and closed-circuit TV cameras. We can use computer communications to check our bank account, order goods via the Internet and even do our work at home.

Computers have changed many people's leisure activities: over 25 per cent of British homes have a PC and this is expected to double over the next few years. Home computers and games consoles offer a new form of entertainment. For example, many home users use software to trace and record their family tree, to plan a journey by accessing route planning and mapping software or access information from an encyclopaedia held on a CD-ROM.

Home use of the Internet is also growing. The use of e-mail to keep in touch with family and friends around the world is becoming increasingly popular. Home users can also book holidays, carry out their personal banking and order goods from a supermarket that will be delivered to their door. Such e-commerce facilities save time.

As with any technology there will be those who want to be ahead of the rest. Others will be frightened of the new technology and try to avoid it. There is the danger that society will split into two: those who have access to the new technology and those who don't.

Activity 1

Working in a group, produce a list of the different ways in which ICT can be used in the home. Then produce a questionnaire for a survey of different households' home access and use of ICT. Produce a report of your findings to share with the rest of the class.

IT and education

Computers are of course common in schools, colleges and universities, and ICT has improved education in other ways. Intelligent tutoring systems enable the computer to give the student information, ask questions, record scores and work at the individual's pace.

Computer communications provide new opportunities for distance learning. Students can send their work to their tutors by e-mail and receive back annotations and comments. Videoconferencing may be used for lectures, enabling two-way communication and discussion. This is particularly useful in remote and under-populated areas. Students in schools or colleges can undertake courses which are not viable to run in their own institution as there are not enough students wishing to study the subject. From a college viewpoint, the use of distance learning increases the potential market for their courses.

Martin is studying to be a teacher, undertaking a Postgraduate Certificate in Education course at the Open University. His tutor, Kate, lives 80 miles away. Martin has to send assignments to Kate regularly. Kate then marks the assignments and sends her comments back to Martin. To help students communicate with their tutor, all the students receive an Apple Mac personal computer at the start of their course. Martin sends his assignments by e-mail.

Martin has found the system very helpful. It has helped him maintain deadlines, as he cannot say that he has lost work in the post and e-mail is date stamped. He can easily send part of his assignments to Kate to make sure he is on the right track. Communication between the student working throughout the day in a school and the tutor working in a college is much easier.

Computer Based Training (CBT) is a sophisticated way of learning with help from ICT. A simulated aircraft cockpit used for pilot training is an example of CBT, which is cost-effective and less dangerous than the real thing. The number of Computer Based Learning packages that are available is growing very fast as modern computers have the processing power and storage capacity to support fast moving and realistic graphics.

The Internet has become a major research tool in schools, colleges and universities.

Many schools are using electronic means for registering their students in class. Different methods are described in Chapter 3. Such methods can provide up to date attendance information brought together from a number of classes that provides a valuable overview of an individual student's attendance as well as providing the opportunity of studying trends in class attendance.

Schools and colleges enter students for public examinations electronically using EDI. Files are sent to the exam boards with details of students and the modules they are entering. Later on the results are sent back to the institution via the Internet where they are imported into a databases and the results are printed out for the students.

Students applying to university through UCAS are likely to do so electronically using the EAS (Electronic Application System). Using this the student fills in an on-line form, entering personal details, course choice details, lists of qualifications gained and examinations to be taken, and a personal statement. Once this has been completed the student's reference is added by a member of staff. The application is then sent electronically to UCAS who forward it to the chosen institutions. (See Figure 4.5.)

Changes in employment patterns

Technological developments since the Industrial Revolution over two hundred years ago have led to changing patterns of employment. The Industrial Revolution led to the building of factories that resulted in a shift in the population from the countryside to the towns.

The development of the computer has also changed patterns of work and it has affected nearly every part of industry and commerce.

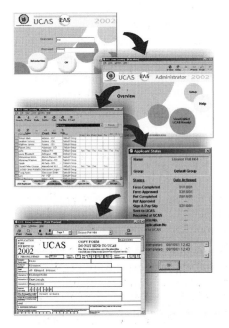

Figure 4.5 UCAS

Some skills have disappeared completely. For example, in the printing industry, typesetting used to be a skilled operation using hot metal. It was performed by print workers who had undergone a seven-year apprenticeship. Now it can be done in the office using a desktop publishing program and a standard PC. This has resulted in greater job flexibility and the breakdown of the traditional demarcation lines between printers and journalists.

Automatic telephone exchanges have cut the number of personnel needed. Football pools checkers are no longer needed as the job can be done automatically.

Some jobs may have changed little, such as gardeners or delivery drivers, but they still may be affected by such inventions as computer-controlled greenhouses or automated stock control. Other jobs such as supermarket checkout operator, bank clerk or secretary have changed considerably. This has usually meant that existing staff have had to retrain to use ICT.

Teleworking

Telecommuting means using ICT to work from home (or another place away from the office). It has been made possible by advances in technology and networking such as fibre optics, faster modems, fax, satellite systems, internal e-mail and teleconferencing. Working from home has been common for a long time in some industries, such as sales representatives, telephone sales and the self-employed. Now telecommuting has extended home working to other industries.

It is now possible for authors, journalists, computer programmers, accountants and word-processor operators to do their work by telecommuting.

Telecommuting means using information technology with communications systems in the form of phone, fax and e-mail to enable people to work away from an office. Some teleworkers work from home while others use small offices sometimes called telecottages. BT estimates that there are around 1.3 million teleworkers in Britain, made up of 650,000 self-employed persons, 150,000 company employees and 500,000 mobile workers such as sales staff.

It is still common for telecommuters to spend part of their working time at the office and part at home. Typically such a person would work at home for three days and be at the office for two. Many companies use a method of 'hot desking', where instead of having a work station for every employee, telecommuters share the use of a number of computers. For the employer, telecommuting saves the cost of office provision such as desks, chairs, space, heating and car parking space.

The pool of available labour for a job is hugely increased if teleworkers are employed. British Airways flight booking takes place in Bombay, proving you don't even need to be in the same continent. Much computer programming and testing for British companies is done in Asia.

There are obvious advantages in telecommuting for the employee such as flexible working hours, avoiding time-wasting rush hour travel as well as the associated costs. Some child care problems can be eased. Telecommuters are no longer tied to living in crowded cities where housing costs are very high but can live in the location of their choice.

However, the lack of the social side of work can be a disadvantage for many employees as many people make friends through their work. Not everyone is good at getting down to work by themselves and some find it hard to be part of a team at such a long distance. Much informal training takes place in a workplace without the participants really knowing about it. There is a real danger for many people that the distinction between work and private life gets blurred and work takes up more and more of their time, effecting the quality of life and producing stress. Telecommuting demands new skills and training is essential if it is to be successful.

	Advantages of teleworking	Disadvantages of teleworking
To the employee	● No need to travel – less stress, saves time and money ● Flexible hours – work around family commitments ● Can live where you like	● Loss of social contact with work associates ● Boundaries between work and home can get blurred leading to stress ● Distractions may make it hard to work at home
To the employer	● Reduced office costs (NB 'hot desking') ● Less absenteeism ● Wider pool for recruitment as location no longer important	● Can be harder to keep track of employees' work ● Potential lack of company spirit
To society	● Less travel leads to reduced pollution ● Family commitments covered in the home ● Reduced concentration of population near employment	● Reduced human interaction at work

In the long term, if telecommuting were to become the way the majority of people worked there would be implications for society. Reduced travel would be environmentally friendly, and could cut pollution and perhaps lead to the end of cities and offices as we know them.

Case Study 3

Building Society Introduces Teleworking

In 1997 the Britannia Building Society launched one of the UK's most ambitious telecommuting projects. The building society implemented a new policy for its text creation department – what used to be called the typing pool.

Dictation of letters is now done over the phone, stored and then transmitted to the teleworker's home. The completed documents are typed at home and submitted by e-mail. Britannia say that the system works really well and it is easy to monitor the work rate, and error rate, of the teleworkers.

Summary

- computers have had an enormous impact on society

- in manufacturing CAD/CAM systems, computer controlled robotic devices and JIT systems are all widely used

- The term 'electronic office' is used to describe how ICT dominates a modern office

- ICT developments effecting commerce include:

- EDI, where commercial data is transferred electronically between companies

  - e-commerce where trade is carried out over the Internet

  - computerised stock control and reordering

- in banks computers are replacing human workers while most people are paid directly into their bank account by computer and an increasing number of other transactions take place electronically

- some people predict a cashless society.

- ICT is widely used in medicine:

- much administrative work is performed by computer

  - expert systems are used as an aid to diagnosis

  - computers are used to monitor thee body functions of patients

  - robots have started to be used in surgery

- ICT is playing a growing role in leisure in the home, where the Internet is used by many people who explore the World Wide Web and send e-mails to friends and family

- in education, ICT supports distance learning, CBT is used extensively in training, the Internet and encyclopaedia CD-ROMs are used for research purposes

- ICT plays a major role in educational administration

- teleworking is the name given to the use of ICT to work from home

The social impact of ICT questions

1. Describe briefly three different systems in which computers are involved in the payment for purchased goods or services. In each case describe clearly the role of the computer. *(3)*

2. Do you think that computers will eventually eliminate the need for cash transactions? Justify your answer. *(2)*

AEB Computing Specimen Paper 2

3. Suggest one benefit and one danger in each of the following applications of computers.

a) The screening of patients before they see a doctor. *(2)*

b) The use of computers to teach school students. *(2)*

c) The use of computer systems to suggest the sentence to be served by a convicted criminal. *(2)*

AEB Computing Specimen Paper 2

4. Briefly describe two social impacts and two organisational impacts commonly identified as a result of introducing computerised information systems into business organisations. *(8)*

NEAB 1998 Paper 1

5. A supermarket chain operates an automatic ordering system between the stores and a central warehouse.

a) State the advantages to the store of using an automatic system. *(2)*

b) Explain one advantage for the supermarket's customers of the store using an automatic stock control system. *(2)*

AQA ICT Module 1 May 2001

6. The use of Information and Communication Technology (ICT) has brought benefits to a number of areas.

For each of the following, state a use of ICT, and describe the benefit that could be gained. Your examples must be different for each case.

a) Education
b) Health
c) The home
d) Offices
e) Manufacturing companies
f) Police *(12)*

AQA ICT Module 1 January 2001

The role of communication systems

All people and organisations need to communicate. Today information technology is a vital tool for all but the smallest businesses. The development of public networks, to which anyone can connect, means that millions of computers can be linked allowing users to communicate with each other, for example by e-mail or by using browser software to access remote web pages. Internet access and e-mail use have increased since the development of other devices such as mobile phones and digital televisions which can also access these services. Text messaging is now a large part of mobile phone use.

Communication technology also includes fax, teletext, palmtop computers and videoconferencing. Data can be transmitted by microwaves (e.g. for wireless computers or communications satellites), electrical currents in wires or light pulses on fibre optic cables.

Data can be transmitted using a **Wide Area Network (WAN)** – a data communications network that covers a relatively broad geographic area and often uses transmission facilities provided by common carriers, such as telephone companies.

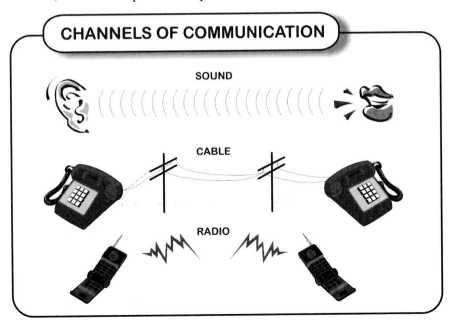

Figure 5.1 Channels of communication

The Internet

The Internet (**Inter**national **Net**work) is a large number of computer networks that can be linked together. Messages and data are sent from the source computer, through a number of other computers until the destination computer is reached.

The **World Wide Web (WWW)** is a vast collection of pages of information held on the Internet. An organisation or individual can set up a web site consisting of stored pages that are made available to other users. Much of the material on the WWW is freely available to anyone. Other pages are password protected and only available to subscribers. Businesses may password protect some of their information pages so that they are available to their employees but not to the general public.

Web sites have a home page which provides access to other pages within the site. Web pages are written in a language called **HTML (hypertext mark-up language).** Web pages can be created and web sites built using web design software such as FrontPage or DreamWeaver. Such software allows a user to set up a structure or map of the web site. He can create, edit or delete pages and set up or edit links between pages to allow easy navigation of the site.

Accessing the Internet

To access the Internet, a home user would normally need

- a computer

- a telephone line which provides the physical connection to a network

- a modem, which is a device for connecting a computer to a telephone line – it converts the digital signal used by the computer into the analogue signal used to transmit data down telephone wires for outgoing messages, and converts the analogue signal into a digital one for incoming messages

- communication software, a browser, a search engine and e-mail software.

A **browser program**, such as Internet Explorer, is needed to view web pages written in HTML in human understandable form. It allows users to retrieve information from the WWW interactively over the Internet. It provides facilities for a user to store the addresses of commonly visited sites as **bookmarks**, store pages and move back through pages recently accessed (see Chapter 14).

The user will need to subscribe to an **Internet Service Provider (ISP)** such as AOL or Freeserve. The ISP will also normally provide an e-mail address to enable the user to send and receive e-mail. The ISP has a host computer that deals with communications and also stores data such as email messages and web pages for the user. The ISP provides a dial-up service that allows the user to link into the host computer and usually offers either

- a local (0845) phone number which allows access to the Internet for the cost of a local call *or*

- a freephone (0800) telephone number to access the Internet if you pay a monthly subscription.

Once set up, the software will dial the number and make the connection to the Internet for the user at the click of a button.

Business users are more likely to use faster digital phone lines **(ISDN)** or broadband connections using fibre optic cables to connect to the Internet. As these lines are digital, a modem is not needed.

Services available

Once connected to the Internet the user can:

- access pages of information on the World Wide Web (WWW), including text, images, sound and occasionally video

- save these pages and images locally for later reference

- leave messages on 'bulletin boards' and join discussion forums

- send and receive e-mail which is stored on the ISP's computer.

The ISP may also provide:

- free web space to set up and edit your own web pages

- additional e-mail address for the user's family

- latest news, weather, TV and radio information

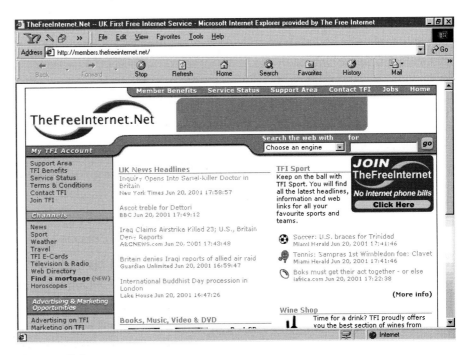

Figure 5.2 The Free Internet

- its own search engine for searching the web

- its own Internet shopping facility including on-line auctions

- bulletin boards newsgroups (special storage space which is used for messages relating to a particular interest group, for example Star Trek, coarse fishing or teaching ICT.

The Free Internet is an ISP that provides news, sport, a wine shop and a search engine (See Figure 5.2).

Search engines such as Yahoo! and Lycos enable users to search the Internet using selected key-words. A search engine is a software program that allows a user to enter a query and will search a very large database to find matching items. This is commonly known as **surfing the net**. The search engines provide a list of links to pages containing those key words. A search on one key-word may provide thousands of links. Finding the right pages can be like finding a needle in a haystack!

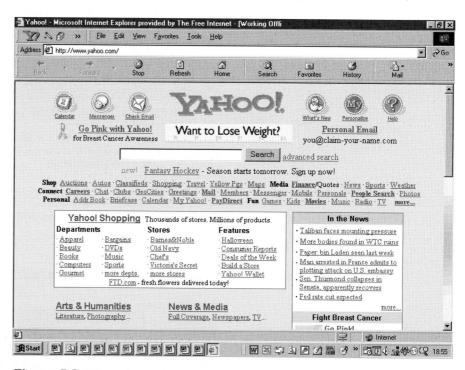

Figure 5.3 Yahoo!

> **Other ways of accessing the Internet** >

Organisations where a number of users are likely to access the Internet are unlikely to find that the use of a modem and dial-up facilities will meet their needs. The user computers are likely to be organised into a local area network which is linked to the Internet via a terminal server. The organisation is likely to have a permanent link to the Internet via an ISDN line.

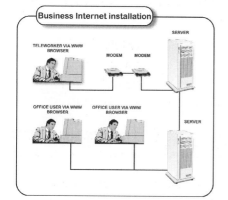

Figure 5.4 Business Internet Installation

Televisions with digital capabilities, either digital TV or a normal TV with a conversion box, allow a user to access the Internet. Some Internet services can be accessed using WAP mobile phones or some new BT phone boxes.

WAP (Wireless Application Protocol) mobile phones provide Internet access and e-mail facilities as well as standard phone activities. WAP is an Internet protocol that has been designed especially for mobile phones. A text-based information service is provided and information is available to the user in the following areas:

- UK News headlines with details of some recent stories
- sports news
- entertainment news together with the facility to search for suitable venues such as clubs or restaurants
- up-to-date share prices
- TV guide
- weather information
- travel information.

Standard e-mail facilities such as reading, replying and forwarding messages are also provided.

What is on the net?

On the 'net' you can find advertisements from companies, government statistics, football club information, details of people's research, pictures of pop stars and the lyrics to their songs. You name it, it's probably there. You can order clothes from catalogues or food and wine from supermarkets using your credit card.

Many newspapers publish editions on the Internet. Users can search to find references to any subject they choose. University researchers put details of their research on to web pages.

Commercial use of the Internet has developed so rapidly that it has spawned a whole new jargon, including such words as e-commerce, e-tailers (electronic retailers). e-business, e-shopping, e-banking, e-learning, e-government, e-procurement and even e-greetings.

Where is a site based?

Internet sites are independent of location. When you visit a site you cannot tell whether the server on which it is stored is in Britain or Bermuda. Internet companies can set up in areas that suit them – not areas close to their market.

For example, First-e is an Internet bank. You've probably seen their advertisements on television. The bank is actually a French bank, based in France, but you wouldn't know that from the advertisements or from visiting the site. Ladbrokes set up their Internet betting site in Gibraltar to avoid paying UK betting tax. As a result of 'off-shore' Internet betting, the Chancellor of the Exchequer had to remove the duty from betting in the UK.

Problems with the Internet (see also Chapter 7)

There is a vast amount of information available on the Internet and it is perhaps inevitable that problems arise. Some concerns include the following points.

- Factually incorrect material is just as easy to store on the Net as correct; however there is a tendency for many people to believe everything that is stored. The Internet contains questionable material such as pornographic pictures.

- The text on some sites is offensive, for example there may be defamatory or racist literature. There is very little that can be done to prevent these offensive sites (see Policing the Internet below)

- Fraud is common on the Internet, for example because sites claiming to sell items are fraudulent or because credit card numbers can be intercepted.

- Workers and students may simply waste time on the Internet, for example visiting sites with games when they should be working. Surveys found that over 60 per cent of visits to pornographic sites were made during company time!

Policing the Internet

The international nature of the Internet and the anonymous nature of the pages make policing very difficult. Some Internet Service Providers try to 'firewall' to filter offensive sites but millions of new pages are being created making the prevention of access difficult and prosecution practically impossible. Software can be installed that will filter out pages that contain any unacceptable words.

Different countries have different laws on such matters. If someone in country A publishes information that is legal in country A but illegal in country B, have they broken the law if someone in country B accesses their pages?

Case Study 1

Internet Fraud

The dangers of using the Internet for buying and selling goods were shown up in the case of 80 year-old Alec Evans. Mr Evans was puzzled when the purchase of more than £200 of goods he did not recognise on his Abbey National Visa card bill. He found out from his local Abbey National branch that the goods had been purchased over the Internet. Yet Mr Evans did not own a computer and had never made a purchase over the Internet.

Abbey National immediately refunded Mr Evans but did not know how the misuse of his card number could have happened, suggesting that he might have been overheard giving out his number by someone who later used the number fraudulently. Visa says that the Internet is not a safe place for financial transactions.

Despite the possible dangers, shopping over the Internet, called e-commerce, is growing rapidly. Dell Computers claim their web site generates $3 million worth of sales every day. Credit card companies say that electronic card fraud is only three per cent of total credit card fraud.

Case Study 2

Dot. com or dot. bomb?

In the late 1990s it seemed that newspapers were full of young entrepreneurs claiming to be dot. com millionaires. All you needed was a good idea and a web site with a memorable name and your future was assured. Or was it?

The future was selling over the Internet and companies like *lastminute.com*, *amazon.com* and *boo.com* were the future. However come the new millennium, shares in dot.coms fell rapidly and several e-tailers went bust. What went wrong?

The Internet expanded very quickly. This growth was mirrored by dot.coms, who needed a lot of money to meet high start-up costs on hardware and storage. Sales prediction were wildly optimistic as shoppers, put off by poor delivery records and fears of fraud, preferred to shop at traditional stores.

Boo.com went bust in summer 2000. Shares in lastminute.com were floated at £3.80 in March 2000. By the end of the year they were down to 78p. Amazon.com had to make 1300 staff redundant in January 2001. On the other hand Tesco.com were very successful doing £5 million worth of business a week over the net.

- Use the Internet to find out what has happened to these dot. com companies. Are they still in business? Is there any particular type of company that has managed to survive?

- Use the Internet to visit three e-commerce sites. Can you find any items that would not be sold in a store near where you live? Are there any products where it does not matter if you cannot see the product before you buy it?

Activity I

This is best carried out in a group of 4 to 5 people.

- Carry out a survey of 20 people each to find out their attitudes and use of the Internet and ecommerce. First prepare a questionnaire. You should include questions that cover:

    1. Use of Internet – time spent and purpose.
    2. Have they used on-line banking or ordered other services or goods on-line?
    3. What is their opinion of the above?

- Share the results of your survey and use suitable software to bring together and present your findings.

The growth of the Internet

The Internet was established in the 1960s for US military and research establishments. It only took off in the 1990s when it expanded greatly due to cheaper computers, faster modems and cheaper subscriptions to ISPs. The number of homes connected to the Internet and the number of web pages expanded very quickly. Companies soon realised that they could advertise over the Internet and use e-mail for fast communications.

As more information appeared on the net, more people decided to go on-line. Businesses, schools, hospitals and councils soon had their own websites and used the Internet and e-mail to do their work.

Intranets

An **intranet** is an internal network, for example within a company, including information pages and electronic mail facilities. It uses the same software as the Internet and can also be connected to the Internet. A company could use this to provide employees with information – for example schedules for the day, stock information or orders due for delivery – as well as internal e-mail.

Case Study 3

Intranet–Electronic Arts

Electronic Arts is a successful software company based in Silicon Valley, California, known for some of the best-selling computer games in the world. EA uses an internal web network, or intranet, to make important information available to its 2300 employees and to enhance teamwork.

EA wanted a system that would allow employees to exchange information in an easy and cost-effective way and could be used by different platforms (types of computer). Using a web browser makes it very easy to access information, which can include text, graphics, sounds and video clips.

Intranet newsgroups can discuss projects over the intranet, enabling EA employees to collaborate to tackle a project regardless of where they are based.

The benefits of Electronic Arts' intranet include:

- information sharing – employees can access all information in one place
- team collaboration – newsgroups enable 'virtual' teams to work together.

- What other benefits is the Intranet likely to bring to Electronic Arts?
- Why would the organisation not simply use the Internet to store all its pages of information?
- What are the benefits of having an Intranet in a school or college?
- List the range of facilities such an Intranet could offer.

> **Extranets**

Companies can extend their corporate intranet systems into an extranet. Extranets are direct network links between two intranets, often between two businesses such as a retail company and their supplier.

The retailer can access their suppliers' stock records, check stock levels and delivery times before ordering directly via the extranet.

Case Study 4

Extranet–Darnton Tiling, Leeds

Darnton carry out tiling work in offices, factories and homes. They need to know if their suppliers can deliver the tiles for a particular job quickly. Darnton has an Internet link with tile manufacturers H&R Johnson Tiles, of Stoke-on-Trent.

Darnton access the manufacturer's home web page and by entering a user name, password and PIN number can find out what is in stock, when new stock will arrive and then place orders. The Internet also allows Darnton to check on tile prices and take advantage of any offers.

- What problems could arise from using such an Extranet?
- Suggest ways of overcoming these problems.

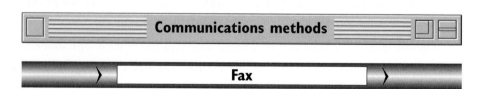

These links can use either the Internet using **encryption** or private leased secure lines. Connections to consumers and to business prospects are directly over the Internet – often encrypted for privacy.

> ## Electronic Data Interchange (EDI)

EDI (Electronic Data Interchange) is a means of transferring information such as invitations to tender, letters, orders and invoices electronically via the telephone network. It allows the computers in one organisation to 'talk' to the computers in their supplier's organisation, regardless of computer manufacturer or software type.

EDI cuts down the paper mountain. Although all large organisations and most smaller ones use computers, it is true that the vast majority are still essentially paper based.

For example, a simple order is raised on one computer, printed, mailed, then received by the supplier who re-keys the details into another computer. The process is expensive, time-consuming and prone to postal delays and errors. EDI changes all that. Acting as a giant, efficient electronic mailbox, it collects the orders directly from one company's computer and sends them to the suppliers' computer. It cuts out printed mailings and removes re-keying, minimises the margin for error and saves days in the processing cycle.

Case Study 5

EDI at Nissan

Nissan's Sunderland plant started production in 1986. Rapid increases in production levels meant that the paperwork generated soon reached large proportions with as many as 15,000 delivery notes each week.

The labour costs of dealing with all this paperwork and the associated mailing costs were excessive. There was also the potential for human error. Following an investigation Nissan started to use EDI in 1989.

Almost immediately there were savings in labour and mailing costs, a shortening of the time for delivery information to reach suppliers and a reduction in the level of human errors. EDI is used to transmit delivery requirements to Nissan's logistics partner, Ryder Distribution Services, who in turn use EDI to send delivery data to Nissan. The volume of mail to suppliers has been reduced by 90–95 per cent.

> ## Communications methods

> ## Fax

Fax is the more common abbreviation for facsimile transmission. A fax machine uses telephone lines to transmit copies of an original document. The fax machine scans the document, encodes the contents and transmits them to another fax machine that decodes the data and automatically prints a copy of the original.

The document is sent as a graphic and therefore takes longer to send than a text file. Obviously you can only send faxes to someone with a fax machine but fax machines are now extremely common in business – in fact it is rare for a business not to advertise its fax number.

E-mail (Electronic mail)

With e-mail software it is very easy to send and receive electronic mail to or from any other person or organisation who has an e-mail address, known as a mailbox, anywhere in the world for the same cost as accessing the Internet (usually a local or free phone call). Most businesses now advertise their e-mail addresses. An e-mail address is usually of the form:

sally.miggins@computerland.co.uk

Addresses are in lower case letters. Words are separated by full stops. The UK at the end is the only indication of the geographical location of this address.

E-mail software contains a text editor that allows a user to prepare a message. These messages are equivalent to traditional letters or memos that are delivered physically. E-mail does not have the facility for a one-to-one conversation as in a telephone conversation.

E-mails can include attachments – computer files that are sent with the e-mail. For example, it is possible to send word-processed documents and images as attachments. The person receiving the e-mail can them store and use these files in the normal way.

Faxes arrive automatically, but not e-mail. E-mails are stored on the Internet Service Provider's computer for the recipient and will be stored there whether or not the recipient's own computer is switched on. Users have to check their mailbox to see if they have any mail. If they forget to check, e-mail isn't very quick! E-mail can also be sent within an organisation on a local area network.

E-mail software can be set up to dial up and check automatically for new mail every hour or so. Some companies save money by storing e-mails sent out until later in the day and sending them all out together at cheap rate. This also means that the e-mails are sent when network traffic is much less. Urgent e-mails can still be sent immediately by giving them 'High Priority'.

Documents sent as e-mail attachments can be loaded by the receiver without the need to re-type them. Journalists send copy to their newspapers by e-mail. The copy can then be imported in the newspapers desktop publishing system. Information sent by fax would have to be re-typed, wasting time and introducing the possibility of errors.

E-mail software such as Microsoft Outlook Express enables users to:

- click on a Reply icon to reply to an e-mail without having to type in the recipient's e-mail address

- forward an e-mail to another e-mail address

- set up an address book of e-mail addresses

- set up a group of several e-mail users to whom the same e-mail can be sent

- set the Priority for an e-mail.

Activity 2

- Use any e-mail software package to carry out the following tasks:

  1. Forward a message received from one person to another.
  2. Set up a group of four or more people and send an e-mail to all members of the group.
  3. Send an e-mail with a word processed document as an attachment.
  4. Send an e-mail that is marked as *priority*.

- Prepare a simple user guide to explain to an inexperienced e-mail user how to carry out each of the four tasks.

Fax modems

There has been **a convergence of technology** in the ICT industry. Computers can be used as televisions, video players, telephones, CD players and fax machines. Fax modems can be used to send faxes as well as e-mails. However, the computer can only receive faxes if it is left turned on in receive mode.

E-mail or fax?

Sending a letter may be too slow. A phone call does not give hard copy. An office in another time zone, for example in Australia or the USA, may be closed. How can a business send a message? What are the advantages of using e-mail and fax?

Internal e-mail is suitable for memos within a business using a LAN network. It is easy to use, involves no paper and costs nothing but in some offices it is replacing personal contact. Employees are sending e-mails to the next door office instead of going to see the recipient personally.

The telephone is still a very useful means of communication, particularly where personal contact is involved and an immediate answer is required.

	Fast?	Hard copy?	Who can receive it?	Can load directly into the computer?	Easy to use?	Cost?
E-mail	Yes, faster than a letter. However, users must check their mailboxes regularly.	Yes	Only those with an e-mail address.	Yes. For example a newspaper can place an article sent by e-mail directly into a desktop publishing program.	The recipient must log in to get it. It is possible to access e-mails from any computer with Internet access.	A subscription to an Internet Service Provider. Calls are either free or at local rate.
Fax	Yes, but a fax sends a message as a picture. Therefore it takes longer than sending an e-mail, which sends the message as text.	Yes	Only those with a fax machine or computer with fax/modem.	No, a newspaper would have to re-type a faxed article.	Yes	Approximately £200 for the machine. Cost of call depends on distance.

Teletext

Teletext is an electronic information service which can be viewed on specially adapted televisions. The user will have to pay about £50 extra for a Teletext TV, which is operated using the TV's remote control. Teletext can be used to view information such as news, weather forecasts, traffic information and sports results provided by the television companies. Teletext is cheap and easy to use and can be read by over 30 million people in the UK, with an average of 18 million people using the service each week. Each page is numbered and transmitted in sequence. When a particular page is requested the viewer must wait until the requested page is next transmitted.

Advantages and disadvantages of using Teletext as an information source

Advantages	Disadvantages
Cheap to use: most households in the UK own a TV set. For a small extra cost teletext provides access to a range of information.	Only a limited amount of information is available.
No need to buy a computer.	Changing from page to page can be slow.
Easy to use; information can be accessed from an armchair using a remote control handset to enter the number of the page required.	The information is only text based; no graphics are displayed.
	Non-interactive.

Viewdata

Viewdata means an interactive information system using a computer, modem and telephone lines. There are a few networks that provide viewdata to the general public. PRESTEL provided by British Telecom was the world's first public viewdata service, starting in 1979. It enabled users to send e-mail and to access and send information, for example home banking or holiday booking, and was a forerunner to the Internet.

Advantages and disadvantages of using Viewdata as an Information source

Advantages	Disadvantages
Interactive: can enter search requests or even order goods	Need to use telephone line, so increased running costs compared to Teletext.
No need to purchase a computer — use TV.	Becoming superseded by the Internet.
Is able to store large quantities of data.	

Videoconferencing

Videoconferencing, sometimes called teleconferencing, means being able to see and interact with people who are geographically apart. Two or more people can be connected to each other. The equipment needed includes a high specification PC, a video camera usually positioned near the monitor, microphone and loud speakers, a high speed modem and high speed communication line, for example an ISDN line.

The use of videoconferencing enables business meetings and interviews to take place avoiding the expense and time of travel. The equipment can be expensive and at present the image quality is still not as good as on television or video but continues to develop as hardware improves.

Comparing videoconferencing with traditional meetings or phone conversations

Advantages	Disadvantages/drawbacks
Saves time and cost of travel for meeting.	Quality of image is often poor.
Useful if a quicker response is required than meeting can allow.	Time delay in transmitted image and sound can interfere with meeting.
Can show physical objects that cannot be done by phone.	High set-up costs.
Can see facial expressions and body language chat are lost in a phone conversation.	Staff need training to make best use of equipment.
	Difficult to control meetings.
	The advantages gained through social interaction and informal meeting can be lost.

Case Study 6

Manager and Mother

Guardian, 31 May 2001

Sarah Higgs has two full-time jobs. Professionally she is global marketing manager for BT Conferencing and domestically she is a mother with a young daughter.

She uses a desktop video system to conduct meetings across different countries and has done so for over a year.

Higgs is able to conduct most of her meetings remotely either through BT's reservation-free 'MeetMe' call-conferencing service or through the videoconferencing system.

Aside from business benefits, Higgs has noted a difference in her work/life balance: her daughter has recently started school and Higgs has been able to work around both picking her daughter up from school and maintaining a demanding international marketing role.

- What hardware does Sarah need to carry out her work?

- In what ways could the Internet be used for marketing her product?

Remote databases

It is now possible to book airline tickets, holidays, train tickets, hotel rooms and even theatre seats from travel agents all over the country. These travel agents can connect to the holiday company's computer using a computer network and find out which holidays are available. When the customer has decided on the holiday, it can be booked electronically and the holiday company's database updated.

This is an example of a remote database. The computer communication enables the customer to know exactly what is on offer and have their booking confirmed immediately.

It is possible for Internet pages to take information from a database. **Active server pages (ASP)** can search an underlying database and show the user selected information, for example when looking for suitable train times.

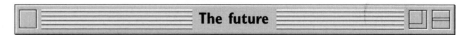

The future

Computer communications have grown greatly in recent years. Processing speeds have increased and transmission methods have improved. More and more people have access to the Internet, either at home, at work, at school, at college or through services in public libraries.

The expansion is likely to continue, as more people have e-mail addresses and there is more information on-line. The UK government wants all its services and local council services to be able to be accessed electronically by 2005. The next generation of mobile phones will be able to include video images. Palmtop computers the size of a pocket calculator will be able to be used as mobile phones and computers. They will have Internet access and e-mail facilities.

With faster Internet access, there will be more opportunity to download large files such as music or videos.

Summary

To access the Internet, a home user would normally need:

- a computer

- a telephone line

- a modem to connect the computer to a telephone line

- a browser, a search engine and e-mail software

The Internet

- enables users can e-mail other users

- provides access to millions of pages of information with excellent research facilities

- can be accessed from anywhere usually for the price of a local phone call

EDI (Electronic Data Interchange) is a method of transferring information electronically between different organisations.

Different communication systems include:

- telephone
- fax
- e-mail
- teleconferencing
- viewdata
- teletext
- videoconferencing

The role of communications systems questions

1. Changes in technology now mean that it is no longer necessary to have a PC to be able to use some Internet services.

   a) Give two devices that can be used instead. *(2)*

   b) Explain why it is possible to send e-mail successfully to someone who has not got his or her PC switched on. *(2)*

   AQA ICT Module 1 May 2001

2. A firm of engineers in the United Kingdom has been given the task of designing a new building in Australia. As one of their methods of communicating with their clients, it has been suggested that the company uses videoconferencing.

   a) Explain what is meant by the term videoconferencing. *(4)*

   b) Give two advantages and two disadvantages to a company of using this approach as compared with non-ICT methods. *(4)*

   AQA ICT Module 1 May 2001

3. The manager of the sales department in a major multi-national company is trying to encourage the use of e-mail amongst her staff for both internal and external communications.

   a) List four functions that the e-mail software should provide. *(4)*

   b) Describe **four** advantages of using e-mail for the department. *(8)*

   c) E-mail is sometimes blamed for causing the computer network to 'slow down'. Describe two measures that can be taken to prevent this problem from occurring. *(4)*

   AQA 2000 Paper 1

4. Give two advantages and one disadvantage of a firm using electronic mail as a method of keeping in touch with its large number of travelling salespersons. *(3)*

   NEAB Specimen Paper 1

5. Give four reasons why is it better for schools and colleges to use e-mail for exam entries *(4)*

6. A large company has introduced a communication system which includes electronic mail. This system will be used both for internal use within the company and for external links to other organisations.

   a) Describe two features of an electronic mail system which may encourage its use for internal communication between colleagues. *(2)*

   b) Contrast the use of an electronic mail system with each of fax and the telephone. *(6)*

   c) Describe two functions the communication system might have, other than the creation and reception of messages. *(4)*

   NEAB 1996 Paper 1

7. Facsimile and computer based electronic mailing systems are different forms of message systems.

   a) For each of these systems, describe two of the facilities offered. *(4)*

   b) Discuss the relative strengths and weaknesses of each of these systems. *(8)*

   NEAB 1998 Paper 1

8. Many businesses now use EDI to communicate with their suppliers

   a) What is meant by EDI? *(3)*

   b) Give four advantages of introducing EDI *(4)*

9. The use of e-mail has increased dramatically over the last five years. This has improved communications both internally within a company, and externally between companies and their suppliers and customers. Describe the facilities of an e-mail software package that you would use to carry out the following tasks efficiently.

   a) Pass on an e-mail message that you have received, in error, from a customer to the sales manager. *(2)*

   b) Inform a group of staff about the time and date of a meeting. *(2)*

   c) Send designs of a new product to the manufacturing department. *(2)*

d) Send an important and urgent message to a supplier.
(2)

AQA ICT Module 1 January 2001

10. A company specialises in organising international conferences for doctors. The company has decided to make use of the Internet for advertising and organising the conferences.

a) State, with reasons, the hardware that the company would need, in addition to their PC and printer, in order to connect to the Internet. *(4)*

b) State the purpose of the following software when used for the Internet:

i)   Browser

ii) Editor

iii) E-mail software *(3)*

c) Explain three potential advantages for this company of using the Internet as opposed to conventional mail/telephone systems. *(6)*

NEAB 1999 Paper 1

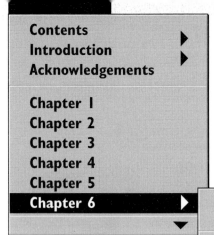

Working with IT

There are many different jobs associated with ICT. Some jobs will involve using ICT only some of the time. Other jobs will involve using ICT most if not all of the time. These include **development personnel** such as systems analysts and programmers who produce new systems, and **operational personnel** such as operators and support staff who keep the system running.

There has been a blurring of definitions as some staff are responsible for development and support, as can be seen from the advertisements below. The Database Administrator post, for example, is involved in investigation, implementation and support.

Senior Technical Test Analyst
Large mobile multimedia operator requires an analyst as part of a team to ensure test strategy, test processes and procedures are implemented through test documentation and execution.

Senior Programmer / TV Company
Working on both broadcast and non-broadcast developments, you must have the ability to apply your skills in a creative way and positively thrive on a diet of broadcast deadlines and achieving the impossible yesterday.

A practical approach to problem solving is essential as is the ability to communicate complex technical concepts to non-technical clients.

Assistant Database Administrator
You will have a degree or equivalent in a computing related subject and have experience of a database administrator environment.

You should have excellent interpersonal and problem solving skills and a flexible and positive approach to managing your own workload and shifting demands.

Corporate ICT Strategist
Playing an influential role, you will be an ambitious professional with a profound understanding of IT and the enthusiasm to rise to new challenges.

You will be a strategic thinker with the ability to grasp complex situations quickly and provide innovative solutions.

With a thorough understanding of current and emerging technologies, you will be an effective communicator, have sound presentation skills and a high degree of drive, motivation and enthusiasm.

Technical Consultants
You will have a background of application design, specification and integration.

You will be able to research and analyse technical requirements at all levels and match these to the needs and constraints of the business.

You will have the capacity to produce creative and cost effective solutions within a complex customer environment. You will be confident with a consultative approach and be an effective communicator with business and technical people.

Figure 6.1 Job advertisements

The range of work in the ICT industry is huge and always changing. Each job will have specific skill and knowledge requirements. There are however personal qualities and general characteristics that are relevant to many fields of work within the ICT industry and it is often the possession of these, as much as specific IT skills, that allows an individual to progress to a senior position.

Figure 6.2 Personal characteristics

Adaptability

The one certainty in the ICT industry is that nothing stays the same! The rate of change in hardware and software performance has been very rapid in the last decade. Skills regularly need updating and old ways of doing things are often abandoned. A successful ICT practitioner must be able to take change in his or her stride, and should have the capacity to adapt easily to new working methods. Much work is project-based and an ICT worker is likely to move between teams, where it is not unusual, as projects overlap, to be a member of more than one team at the same time.

An openness to new ideas is essential together with a real enthusiasm for learning. Sometimes courses are available for updating skills, but the most effective employee is able to acquire new skills in a variety of ways.

Good communications

The computer boffin, who rarely speaks to anyone but sits alone all day hunched over a computer screen, is a common public image of a person in the ICT industry. In fact there are few jobs that allow such isolation. Most require some communication with others – both oral and written.

Written

At all stages of the system life cycle written reports are produced – feasibility studies, specifications, maintenance and user guides as well as progress reports for management. A programmer needs to produce written documentation for his programs so that, when the program needs to be modified at a later date, other programmers can gain a clear understanding of what the program does. The ICT professional will need to be competent in writing a variety of documents in good clear language, in a style and using a level of technical detail that is appropriate for the audience.

Oral

When working with users of a system, ICT professionals need to be able to discuss problems and provide solutions in a clear, jargon-free and friendly way. Considerable time may be spent in face-to-face discussion with users who have little knowledge of the technicalities of ICT systems. If an ICT support worker does not have good communication skills, he is likely to confuse a user when he tries to help them solve their problems. Care must be taken not to use language that will antagonise, patronise or confuse in such circumstances.

A system analyst will need to be a good listener so that a user's requirements are fully taken into account. Many new systems have proved to be unsatisfactory because the ICT specialist produced what he thought the user should want rather than what was actually needed. The analyst must also be able to explain new ideas clearly.

Integrity

An employee in the ICT industry must be trustworthy. An individual might have access to sensitive information and it is important that they can be relied upon not to misuse it. Many organisations will have a code of conduct which lays out rules of behaviour for employees. (Codes of conduct are studied in A2). Such a code of conduct could include the following:

An employee should not:

- spend work time sending personal e-mails
- let anyone else know his/her password
- try to access data not relevant to his/her work
- install unlicenced software on his/her workstation
- divulge customer information.

Personal organisation

Many ICT jobs require the employee to work on a number of different tasks concurrently, so the ability to organise one's time and to prioritise tasks is vital. It is important that deadlines are met and work is not left to the last minute.

Team working and getting on with a variety of people

The ability to work effectively in a team is essential for nearly all ICT practitioners. A good team member is sensitive to the needs of other members, reliable, supportive and cooperative. In a strong team, ideas are freely shared and members build on the strengths of others while supporting their weaknesses. ICT workers are likely to need to interact with a range of people in an open and non-threatening way. When a member of an ICT support team goes to a user to help with a problem the user may be angry or frustrated. The support worker must act tactfully, politely and supportively.

Thoroughness and attention to detail

There are many ICT jobs that require precision and a detailed approach. For example, if a programmer does not follow the specification exactly then the resulting program may have unpredictable effects when implemented. Incorrect data entry will lead to incorrect information.

Creative flair

The ability to think innovatively and come up with new ideas is needed in many ICT jobs. An outstanding programmer needs to have more than the necessary technical language skills. To solve some problems, an effective programmer will need to think of new ways of doing things. Similarly, if a web designer is to create exciting and effective sites, she will need to possess a strong visual sense and spatial awareness as well as sound technical knowledge.

Analytical approach to problem solving

An ability to approach problem solving in a systematic and logical way is essential for certain ICT roles such as programming, network management and systems analysis. Many organisations require new employees to take aptitude tests to demonstrate that they can take this approach.

Ability to manage pressure

There are likely to be many situations when an ICT professional is under pressure: a deadline could be looming, a program might not be performing as it should, a crucial hardware device could fail or a user could come up with unexpected and urgent demands. The pressures might mean that the employee has to work very long hours for a period of time.

Willingness to work flexible hours

In certain jobs it is essential that the employee is able to work flexible hours. Many systems for multi-national organisations are worldwide and different offices will be in very different time zones. An employee providing software support could be working in London, whilst the users are located in Los Angeles. When a new system is introduced it is likely that he will be required to work late into the night so that he is available to answer queries that occur during the Los Angeles working day. An IT support worker is often required to be available 'on call'. This means that for certain hours, when the employee is not in the office, he must be available to be contacted by pager to deal with queries. These could occur in the middle of the night.

General characteristics

A good level of literacy and numeracy will be needed in practically all ICT roles. Common sense is very important too!

Skills and knowledge

Each job will have its own technical skills requirements. These might relate to the characteristics of specific hardware, the use of a range of facilities offered by particular software or perhaps the knowledge of a given programming language. Such skill requirements are not static and an ICT professional will need to regularly update their skills.

A knowledge of general business practice is essential for many ICT roles as well as a detailed understanding of the specific industry, such as banking, retail or education.

Activity 1

Imagine that you are preparing your CV to use in applications for jobs in the ICT industry.

- Prepare a list of your ICT skills and knowledge.

- Itemise the personal qualities that you possess which would show that you are suitable for employment. Back up each quality that you list with an example to support your claim.

Activity 2

Use the Internet to search for job advertisements in the following fields:

- Database Administrator
- Data entry Clerk
- Network Administrator
- Web Designer
- Programmer
- Systems Analyst
- User Support

Draw up a table with the following headings to display your findings:

Job title	Skills and experience	Personal qualities

For each of the ict professionals whose jobs are summarised below in the four case studies, describe the personal characteristic that they need to be successful and explain why each is needed.

Case Study 1

Jeremy

Jeremy is the senior IT technician in a FE college and has three technicians working to him. The normal working day is 8.30 a.m. to 5 p.m., but once every four weeks he has to stay until 8p.m. on a Monday to Thursday as well as cover on Saturday. The next week he will have Monday off. Some tasks, such as adding a network printer to a classroom, which involves installing the printer driver in all 20 computers, can overrun at the end of the day. The task cannot be left as the computers will be needed by a class at 9 a.m. the next day.

Jeremy's job is varied; when he gets to college his first task is to check that nothing has gone wrong overnight and that the back-up procedures took place satisfactorily. Part of the day will be spent staffing the help desk. In between calls he will get on with other tasks such as monitoring the network usage, studying the documentation for new software or preparing material for a staff training session.

At the help desk, all user problems are logged on a database and when a problem is solved the resolution is added. Often a problem will require Jeremy to go to the user to resolve the fault on site. There are over 300 members of staff with vastly differing IT skills; Jeremy has to make sure that the explanation and help he gives are appropriate to that user and that he neither goes over their head on the one hand nor patronises them on the other. For example, it is important that someone who reports a printer fault is not made to feel stupid when it turns out that they have failed to check that the mains lead is attached. On the other hand, they need to be made aware so that they won't call the help desk for the same problem again!

Much of Jeremy's time is spent on housekeeping and other network jobs. To enable him to do this he has full access rights to the entire network.

Case Study 2

Sophie

Sophie works as a Project Manager in the central IT department of a high street bank.

After gathering user requirements, she designs the system solution and specifies the requirement for developers on her team. She manages the implementation of the solution including the physical release of software. Knowledge of the business area and the system enables her to design the best solution for the users. She always has to consider the users' role and how she can best solve their problems.

The initial stages of the projects involve requirements gathering and the production of specifications with the users. This liaison will continue throughout the project in informal meetings with the users during development and implementation.

On a more formal level, representatives from each area involved in the project meet on a monthly basis, where Sophie reports on the project's progress and the major issues are discussed.

IT in the banking world work tends to be driven by deadlines. As a project nears the time when it needs to be delivered it is often necessary to undertake evening and weekend work. The system that Sophie works on has global coverage so different time zones often require late and early conference meetings. She currently travels to Paris and New York as her two main projects are based there.

She often has new requirements that arise that have not been planned for. This has happened recently where a new project had been prioritised by someone very senior in the bank. This causes difficulties because she either has to manage to fit it in or tell other users that their deadlines cannot be met. This can be frustrating for the team who wants to do a good job and get the project finished.

Her team is currently providing post-implementation support for a completed project and providing hand-over to those responsible for day-to-day support. She is also managing two further projects in parallel which are at different stages in the project cycle, as well assisting on two other projects for which her specific expertise is required.

Case Study 3

Nathan

Nathan is a technical consultant for a business specialing in producing software for conducting e-business over the Internet and using wireless devices. He has to use his extensive technical knowledge to provide a detailed solution for a user's application and hardware. He needs to give this in terms that the user will understand.

When Nathan first meets a customer, they can be in a state of panic, as he is often called in to resolve a live problem that is affecting current service. To reassure them that he can solve the problem, he needs to ask questions before the meeting and carry out some background research. He has to ask clear and concise questions and not get lost in too much detail. He needs to develop a rapport with a client and must establish some common ground and avoid disagreement.

Meeting with clients is only part of Nathan's job. He also has to oversee the development of new software, resolve performance issues, discuss hardware requirements and provide plans for the next release of software. Many of the tasks he needs to complete stretch over a number of weeks. Each week he has to decide how much attention is needed by each task. In general he works on his own with his client.

Nathan finds consulting an exciting and diverse career but does not enjoy the travelling that it involves. His daily travel varies from an hour on the train to two hours in a car, with the additional complication that his location from week to week, and sometimes day to day, is not known until the last minute.

Case Study 4

Mike

Mike is the head of an IT department within an International Investment House. Systems within the organisation are global, being in use in many different countries.

Mike's team of 25 systems analysts and developers produces financial software for clients within the organisation. Mike is responsible for the career development, motivation and well-being of his team.

At the start of a project, Mike needs to write a formal, business specification that must be agreed with the client before he hands over to one of the analysts within his team to produce the technical specification for the system.

Mike controls the release of all new software produced by his team. Any changes of code need his approval before they can be implemented. He has access to all the systems within his control and is accountable for maintaining security registers that track exactly what changes were made by which team member. All the original source code and specifications are stored on a secure disk area to which only Mike has access.

Mike regularly works a 12-hour day and is on call 24 hours a day, 7 days a week. He is issued with a pager, a mobile phone and a laptop computer that allows him to access systems remotely in order to problem solve. On average, he is paged three or four times a week. When all the systems had to be converted to cater for EMU (European Monetary Union), Mike and his team had to be in work for 76 hours continuously.

Summary

As well as requiring specific ICT skills, professionals in the industry are likely to need some or all of the following personal qualities:

- adaptability
- good communications: written and oral
- integrity
- personal organisation
- team working and getting on with a variety of people
- thoroughness and attention to detail
- creative flair
- analytical approach to problem solving
- ability to manage pressure
- willingness to work flexible hours

Working with IT questions

1. Professional progression within the IT industry requires more than just technical skills. Give three other necessary qualities and explain why they are important. *(6)*

   NEAB 1997 Paper 1

2. A software house is advertising for an analyst programmer to join one of their development teams. State four personal qualities that the company should be looking for in the applicants. *(4)*

   NEAB 1999 Paper 1

3. A company is recruiting a new member of staff for their IT support desk. The head of personnel asks the manager of the support desk what personal qualities the new employee must have in order to carry out the job effectively. State, with reasons, four personal qualities that the manager would want a new employee to have. *(8)*

   AQA ICT Module 1 Jan 2001

4. An advertisement for an IT support worker to join the PC support team in a company specifies that the applicant must be 'willing to work flexible hours, be able to communicate well orally, have good written skills and get on well with a wide variety of people'.

   Explain, giving examples, why each of these characteristics is important for someone working in such a role. *(8)*

   AQA ICT Module 1 May 2001

Information systems, malpractice and crime

What is computer crime?

Computer crime is any criminal act that has been committed using a computer as the principal tool. As the role of computers in society has increased, opportunities for crime have been created that never existed before.

Computer crime can take the form of the theft of money (for example, the transfer of payments to the wrong accounts), the theft of information (for example, from files or databases), the theft of goods (by their diversion to the wrong destination) or malicious vandalism (for example, destruction of data or introducing viruses).

What is malpractice?

Malpractice is defined as negligent or improper professional behaviour. It occurs when employees, although not breaking the law, perform acts that go against their professional code of conduct and, intentionally or unintentionally, cause harm to their organisation or clients. An employee who carelessly leaves his workstation logged on, or divulges his password to others, could be enabling unauthorised access to data. Excessive use of a computer at work for personal use by an employee could be malpractice

Why is computer crime on the increase?

The rapid spread of personal computers and particularly distributed processing and WANs, has made information held on computer more vulnerable.

Every single one of the top 100 companies in the FTSE index has been targeted or actually burgled by the new computer criminals. The British police have evidence of 70,000 cases where systems have been penetrated and information extracted. One enquiry revealed three hackers had been involved in making 15,000 extractions from systems.

The arrival of Automated Teller Machines provides a good example of how a new technological device creates new opportunities for fraudulent activity. In the 'phantom withdrawals' scandal of 1992, British banks and building societies were sued by hundreds of customers who claimed they had been wrongly debited throughout the 1980s for withdrawals they did not make. The banks claimed that the customers must have withdrawn the money and that phantom withdrawals from their machines were 'impossible'.

In another case a criminal gang rented a shop, made it look like a bank and installed a fake cash machine. The machine did not issue any money (saying it was out of order) but copied the magnetic strip on the back of the card and stored the card's PIN-number. The gang then made duplicate cards and used real cash machines to steal money.

The advent of the mobile phone has led to another computer crime – cloning the chip inside the phone so that you can use your phone while the charge appears on someone else's bill.

Banking security experts in the USA estimate that an average bank robbery nets $1900 and the perpetrator gets prosecuted 82 per cent of the time. With a computer fraud, the proceeds are nearer to $250,000 and less than two per cent of the offenders get prosecuted.

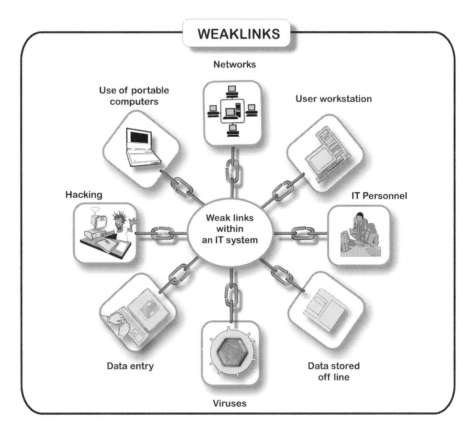

Figure 7.1 Weaknesses of IT systems

Weak Points within an IT system

The weak points within an IT system are associated with hardware, software or people. These days, IT systems are used to carry out all the core functions of a business. They are used for managing finances, stock and many other valuable resources within an organisation. Although systems are under threat from people outside the organisation, employees working within can have the opportunity and skills that allow them to commit fraud. Such a fraud could involve an employee instructing the system to organise the transfer of goods to a particular location where he could arrange for them to be collected and sold for his own profit. Data is vulnerable when it is being entered into a system, stored either on-line or off-line, being processed or transmitted over a network.

Data entry

Data can be fraudulently entered into the system with criminal intent. A corrupt data entry clerk could purposely enter the wrong account number for a transaction so that an unsuspecting account holder is debited. An employee who was operating a system that photographed students and produced college identification cards, was caught accepting bribes from students to input false dates of birth to be printed on to the card.

Case Study 1

Bank Fraud

A woman who opened a bank account using false information, saying that she expected to receive her divorce settlement shortly, carried out a more elaborate fraud. She later returned to the bank and surreptitiously removed all the paying-in slips (used by customers to pay money into their accounts) and replaced them with paying-in slips that she had had specially printed. These paying in slips were exactly the same except they had her account number printed at the bottom in MICR characters – just like the paying-in slips at the back of a chequebook.

When reading paying-in slips, the computer looks for the MICR numbers. If there are none, the operator has to type in the bank account number given. If there are MICR numbers, the information is automatically read and not checked. Money paid in with the fake paying-in slips was paid directly into the woman's account. Customers did not notice any errors until they checked their bank statements. By this time the woman had withdrawn over $150,000 in cash from her 'divorce settlement', disappeared and was never seen again.

User workstations

User personal computers are particularly vulnerable, especially if they are attached to a network. If unauthorised users can gain access to the system they could be able to retrieve or alter data. This could be achieved if a computer user were to leave her computer unattended whilst logged on to a system, with no form of protection.

Use of portable computers

The use of laptop and palmtop computers produces risks whenever sensitive data is being stored. Such devices are likely to be removed from an organisation's premises, where security can be controlled. Access to the company network is often made via a modem using the public telephone system, perhaps from home or in a hotel room.

Data stored off-line

Data that is stored off-line, on floppy disk, cartridge or other devices is vulnerable to loss or theft.

Tape and disk stores should be kept locked when left unattended. There should be formal clerical systems in place so that details are recorded whenever files leave the store. The filing and recording system should be maintained rigourously to ensure that files are not mislaid.

Viruses

A virus is a program that is written with the sole purpose of infecting computer systems. Most viruses cause damage to files that are stored on the computer's hard disk. A type of virus program called a **worm** can reproduce itself. A virus on a hard disk of an infected computer can reproduce itself onto a floppy disk. When the floppy disk is used on a second computer, the virus copies itself on to this computer's hard disk. This copying is hidden and automatic and the user is usually unaware of the existence of the virus – until something goes wrong. Another form of virus is the Trojan Horse, a destructive program that passes itself off as an innocent program.

Thousands of viruses exist with their damage varying from the trivial to the disastrous. The most common virus in the world, *Form*, makes the speaker beep when you press a key on the 18th day of each month. It does not damage the hard disk. The *Jerusalem* virus is more serious. It deletes a program you try to run on Friday 13th. The *Dark Avenger* virus is very dangerous as it corrupts the hard disk and back-up copies. Viruses have caused considerable harm to information systems; whole systems have become unusable and complete files of data have been lost.

Hacking

Unauthorised access refers to the use of a computer by a person who has not been given permission to do so. Anyone who gains unauthorised access is known as a **hacker**. Hacking is often achieved via

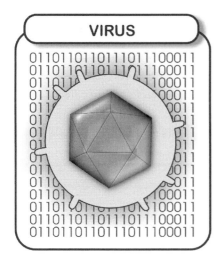

VIRUS

01101101101110111100011
01101101101111011100011
011011.....11100011
011.....00011
011.....0011
01.....011
01.....011
01.....011
01.....011
011.....011
011.....011
01110.....011
01101.....00011
01101.....100011
01101101101110111100011
01101101101110111100011

Figure 7.2 Virus

Activity 1

- Research the latest viruses and write a report describing five current viruses. Explain the nature of each virus and how it is spread.

- Produce an anti-virus guide that provides a user with a set of tips that will help them to ensure their computer does not become infected.

telecommunications links. Many hackers have no specific fraudulent intent but enjoy the challenge of breaking into a system. Although they may have considerable technical expertise, it is possible obtain guides for hackers by down-loading from the Internet. These allow less expert users to become hackers. In some instances the hacker's purpose in accessing the system could be to commit fraud, to steal commercially valuable data or even to cause damage to data that will have serious consequences for the company whose system it is.

Case Study 2

A Hacker Seizes Control of a US Battleship

In a case frighteningly similar to the film 'War Games,' a young hacker gained unauthorised control of the US Navy's Atlantic Fleet via the Internet in 1995. The hacker was a young US air force captain, who showed the US Navy how easy it was to hack into their control systems.

With the sceptical 'top brass' from the Pentagon watching, the young officer used a PC and a modem to control the US Navy's warships. He had no special insider knowledge but was a computer whizz-kid – just the sort of person the Pentagon most want to keep out.

Within a few minutes of logging on to the Net, the computer screen announced 'Control is complete'. Meanwhile, out at sea, the ship's captain had no idea that the hacker had command of his multi-million-dollar warship. Other ships' controls were also taken over.

Networks

Data being transmitted over a network is particularly vulnerable to external threat. The risk of unauthorised access increases when data is transferred over a WAN using public communication links. A line can be tapped to allow eavesdropping of the signal being transmitted along the link. This has been recognised as a real problem for Internet users. It is thought that public worries over the risks to security of using a credit card to order goods via the Internet have increased.

Internet

The rapid growth in the use of the Internet, for advertising, selling goods and for communication, has made many information systems vulnerable to attack.

IT personnel

Security procedures are only as good as the people using and enforcing them. A high percentage of breaches of a company's security are made by its own employees. Disgruntled, dishonest and greedy employees can pose a big threat to an organsation as they have easy access to the information system. They may be seeking personal gain and it is not unknown for employees to be bribed to provide information to a rival. Data may be altered or erased to sabotage the efforts of a company. Information about a business may be of great value to a competitor. Industrial espionage does exist in the cut-throat competitive world of big business.

One computer fraud involves transferring monetary sums to a fictitious account. There is a story of an American bank employee who rounded every interest calculation down to the nearest cent. All the odd fractions of a cent left over went into his account. It added up to millions but no one missed the odd half cent. He would have got away with it if he hadn't started spending the money. His jealous colleagues checked his bank account.

Methods of protection

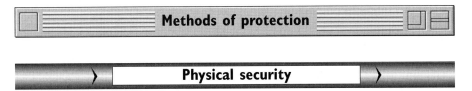

Physical security

The most obvious way to protect access to data is to lock the door to any computer installation. The lock can be operated by a conventional key, a 'swipe' card or a code number typed into a key-pad. Of course it is essential that any such code must be kept secret. Staff should not lend keys or swipe cards to anybody else. Locks activated by voice recognition or fingerprint comparison offer alternative, but expensive, stronger methods of security. (See section on Biometrics on page 79.)

Additional physical security measures include computer keyboard locks, closed circuit television cameras, security staff and alarm systems. Passive infra-red alarm systems to detect body heat and movement are commonly used, as they are reliable and inexpensive.

Computer systems with terminals at remote sites are a weak link in any system as access to them could provide an intruder with access to the whole system. It is essential therefore that such terminals be fully protected. Computer workstations should be logged off whenever they are not in use, especially if the user is away from his desk. Disk and tape libraries also need to be protected, otherwise it would be possible for a thief to take file media to another computer with compatible hardware and software.

Staff and authorised visitors should wear identity cards, which cannot be copied and should contain a photograph. These are effective and cheap. These security methods are only effective if the supporting administrative procedures are properly adhered to, for example doors must not be left unlocked and security staff should check identity cards.

The security measures used by an organisation will reflect the value of the data stored and the consequences of data loss, alteration or theft. Financial institutions like banks need to have the very highest levels of security to prevent fraud.

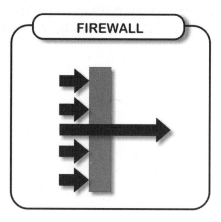

Figure 7.3 Firewall

Firewall

A firewall is an intelligent device that is used to prevent unauthorised access to an organisation's network. The firewall is placed between the network file server and the external network, often the Internet. The firewall checks all messages sent to the file server from outside and filters the contents. A firewall can be used to protect the business's computers from intruders. It is a single security point through which all traffic must pass. The firewall can use passwords to control traffic and can also log details of all attempted access to the site. A firewall allows organisations to manage and control access easily, greatly reducing the risk of network break-in and the destruction or theft of data.

Virus protection: prevention, detection and repair

The risk of getting a virus can be reduced by not allowing users to bring their home floppy disks to use on the system, or to take the company's disks home to use on their own PC. Systems can be set up to only allow specially formatted disks, so that users cannot use their home computer disks. Floppy disks should be write-protected whenever possible.

Viruses can be detected and damage repaired using Anti-Virus Toolkit software. This sort of software is widely available and can detect and repair thousands of viruses. The software is set up to run in the background whenever the computer is on. If an infected disk is placed in the computer's drive, a warning message appears on the screen. The software is usually able to remove the virus. Updates of this software are produced every month as new viruses are detected.

Staff may have to put floppy disks into a 'sheep-dip' workstation before use on the organisation's other computers. This work-station is fitted with the latest virus detectors. Organisations may also use PCs without floppy disk drives, or install disk locks to avoid virus problems.

Identification of users

To make sure that unauthorised users do not access a networked system, all authorised users must be able to be recognised. A common way to do this involves every user being allocated a user identification that is backed-up by a secret password. Only when the identification number and password are keyed in, in response to a series of questions displayed on the screen, is the user able to use any of the software or data files. The user will be able to access a part of the software and data held on the system. There will be different levels of permitted access for different users depending on their needs.

Levels of permitted access

Users can be given one of several different access rights to a system. The rights can be different for different parts of the system.

Possible access rights include:

Full rights –	a user can carry out any action on the file or data
Read only –	the data can be accessed to be viewed or printed, but not altered in any way
Read and write –	the user can read or create new data records
Amend –	the user can change the data held in a record
Delete –	the user can delete a whole record
No access –	the user is barred from any form of access to the data

A student user of a college network has full rights to their own file space. No other user, except the network supervisor, has any rights to the area. There are areas to which the student has read-only access. Here teachers, who have full rights to the area, can save (write) documents for student use. Another area of file space can be used by teachers to store data but a student will have no access rights.

Activity 2

A holiday company, Sun Fun, makes bookings only through a number of travel agencies around the UK. The company produces a brochure twice a year and up-to-date information about all its holidays, including destinations and dates and prices, is available to any customer on Sun Fun's web site.

Registered travel agencies can check availability on specific holidays and make bookings for clients. They can make minor modifications to their customers' bookings at a later date if necessary.

At Sun Fun head office lists are produced that contain information of all clients on particular holidays. Details of new holidays are added and information on completed holidays is archived.

- Customers using the web site can view details of Sun Fun's holidays. What access rights do they have to this data?

- List some categories of data to which customers have no access rights.

- Explain in detail the access rights that would be given to a specific travel agency, making it clear what categories of data each right applies to.

- Who, if anyone, would be given full access rights to the system? Justify your answer.

Some areas of the disk will be write-protected for everyone except the network manager. For example, only the network manager should be allowed to install new and delete old software; add new and delete old users; delete old files, copy files and set files to read-only or read-write. To access very sensitive information, users will need to know several passwords.

Activity 3

Mr Blyth runs a shop in a chain of newsagents, FastNews. The in shop system has two linked computers, one on the counter and one in the back office. The shop assistants use the counter computer to inform customers of the amount that is owed and enter details of any payments made. Any details of changes in orders, cancellations due to holidays or closure of accounts are recorded in a black book. New customers fill in a form with details of their address and paper requirements.

Mr Blyth uses the computer in the back office to produce round lists for the paper girls and boys. Every night he uses the data from the black book to update the computer records as well as create records for the new customers. Once a month he runs a program that produces sales statistics that he sends to the head office of FastNews.

If any problems occur with the system, FastNews send their technician to the shop to sort it out.

- Describe the user rights that for each of the three categories of user: Mr Blyth, the shop assistants and the FastNews technician. You may need to consider different categories of data.

It is important that users follow strict procedures to maintain security. Network operating system software can be set to force each user to change his password after a set time has elapsed. A user should only be allowed to mis-type his password a few times (typically three). If by the third attempt the correct password has not been entered, then the user should be denied access and an entry made in the network log. Users should be encouraged to choose a password with care. An ideal password is a random collection of letters and numbers that has no meaning. Meaningful words and names should always be avoided.

A network access log can be kept. This keeps a records of the usernames of all users of the network, which station they have used, the time they logged on, the time they logged off, which programs they have used and which files they have created or accessed.

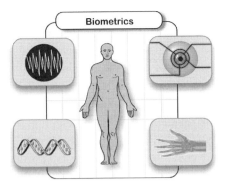

The use of identification codes backed-up by passwords can be vulnerable to hacking. Other, more secure, methods that identify a user have been developed although most are at present very expensive.

Biometrics

Biometrics is the name given to techniques that convert a human characteristic such as a fingerprint into a digital form that can be stored in a computer. Such human characteristics are unique. When an individual tries to access the system his fingerprint is scanned, converted into digital form and compared with the saved version to verify identification. Currently the face, the shape of the hand, the eye and the voice are actually used for identification as well as the fingerprint.

The face is the simplest characteristic to record but unfortunately it changes over time and is therefore only useful in limited cases. A face recognition system is used at an airport in Malaysia to identify passengers booking in hand luggage. A digital photograph is taken of each passenger that is stored on a reusable smart card. A similar card is placed on the luggage and the two are matched when they arrive at the departure gate.

The first use of biometric technology was made in the control of access to buildings. It is now being tested in applications such as e-commerce, banking transactions and data security.

IBM has developed an experimental system, FastGate, which uses a hand, voice or fingerprint to ease business passengers through passport control. This system identifies passengers by comparing their fingerprints, voice patterns or palm prints with a digitised record stored on a central database.

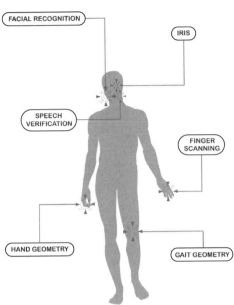

Figure 7.4 Biometrics

Case Study 3

The Eyes Have It
Guardian Unlimited 1999

Soon a cash machine will look you right in the eye before it decides if you can have your money. The system works by photographing your eye and converting the patterns on the iris – the coloured area around the pupil – into a sort of barcode that's as unique to each individual as a set of fingerprints. Other machines may recognise your face, your voice, your signature, your smell, or the shape of your hand. All these are part of a rapidly emerging market for computerised 'biometrics': the identification of individuals by something that is physical, and unforgettable.

Earlier this month, Bank United installed America's first Iris Recognition Automated Teller Machines in Texas. In the world's first public trial, the Nationwide installed ATMs using the same iris recognition technology. Fingerprint verification has been used in cash machines in South Africa since 1997, while in Spain, the social services department is paying benefits using finger-imaging system with smart cards and 650 ATMs.

Biometrics are not just being used for financial transactions. Hand scanners have been installed to control access to thousands of places including the Olympic Village in Atlanta in 1996 and a Los Angeles sperm bank. Since 1993, they have been used in some US airports as part of a project, which enables participants to clear US immigration in about 20 seconds.

continued ...

Case Study 3 *continued*

Voice recognition systems are being used for applications such as telephone banking, where the pioneers include America's Chase Manhattan Bank. Dynamic signature verification – where a touch-sensitive pad is used to capture the speed and style of a signature, not just its appearance – has been tried in the UK for people claiming employment benefits in Liverpool and in the canteen in Pentonville prison. Face recognition systems have been used to identify troublemakers in the crowd at Watford FC, and to secure the Pentagon's computer network. In the UK, Mastiff Electronic Systems has reportedly been working on a smell-based system.

PC users can buy a face recognition program for Microsoft Windows, which allows only people it knows to use the computer. If anyone else sits at your desk while you're out, the package will keep a snapshot of them on the PC's hard drive.

There are three levels of security based on what someone carries, what he or she knows, and what he or she is. The last of these is biometrics. At the moment, the banking industry typically relies on the first two forms of authentication: a cheque card (what someone carries) and a four-digit PIN or personal identification number (what someone knows). They'd like to replace the PIN or password with a biometric measurement.

○ Produce a table on the use of biometric methods of identification that are described in this article.

Biometric method	When used

○ Research further uses and add to the table

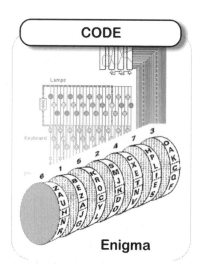

CODE

Enigma

Figure 7.5 Coding

Data encryption methods are used to protect important and confidential information when it is stored or during transmission from one device to another. Encryption is a method of scrambling or coding a message so that someone who intercepts it, for example by breaking into a cable, cannot understand or change the message.

An example of the use of encryption occurs in the banks' Electronic Funds Transfer (EFT) system. Banks and other financial institutions transfer very large amounts of money electronically. These transfers are protected by the use of data encryption techniques.

The simplest of all the methods of encrypting data uses a translation table. Each character is replaced by a code character from a table.

However this method is relatively easy for code breakers to decipher. More sophisticated methods use two or more tables. An example of this method might use translation table 'A' on all of the even bytes and translation table 'B' on all of the odd bytes. The use of more than one translation table makes code-breaking relatively difficult.

Even more sophisticated methods exist based on patterns, random numbers and the use of a key to send data in a different order. Combinations of more than one encryption method make it even more difficult for code-breakers to determine how to decipher encrypted data.

Protecting against internal crime

Employees working in sensitive areas must be totally reliable. They will often need to be vetted before appointment. Strict codes of conduct exist for employees and anyone found to be in breach of these regulations is likely to be dismissed.

The use of an audit trail can enable irregular activities to be detected. An audit trail is an automatic record made of any transactions carried out by a computer system. Whenever a file is updated, or a record deleted, an entry will appear in the trail. The means that a record will be kept of any fraudulent or malicious transactions or deletions that are made.

If possible, the different stages involved in carrying out a transaction should be divided up so that no one person is responsible for the whole process. This method is also used when a program is being written: no one programmer will be allowed to write the whole program so that producing code that carries out fraudulent transaction will be very hard to achieve without the involvement of other programmers.

The Internet: issues

Some people feel that the Internet poses a threat to society itself. The Archbishop of York has warned that computer 'wizardry' is in danger of creating a 'society without a soul….the danger is in having all this wizardry in individual homes which people never leave and where there is, as a result, no social interaction.'

It is widely agreed that access to information brings power in today's world. The access gap is growing between the 'haves' and the 'have nots'. The citizens of the developing world, already disadvantaged through lack of resources, are falling further behind as few will have access to the Internet.

The global nature of the Internet, which provides open access to millions of users worldwide, makes policing it very difficult. There is no overall body governing the running of the Internet. The international nature of the Internet means that information is crossing national boundaries all the time; different countries have different laws surrounding the use of the Internet. It is hard to trace the ownership of sites and new ones are springing up all the time.

There is no regulatory body which determines who can publish material on the Internet. It is easy for anyone to set up a web site and post material there, making any information they like easily available to millions of web users. There are no checks made on the accuracy of the data that is stored.

Many people tend to assume that because some information is on the web it must be correct. A vet has noticed an alarming trend of pet owners using information from the Internet to diagnose their pets ailments for themselves. Owners are using sites where they can input symptoms and get a diagnosis. Unfortunately the information is often wrong and potentially risks the health of the animal. The vet was contacted by an owner who had incorrectly diagnosed her cat as suffering from a thyroid problem. The treatment that the owner gave from the advice given on the web nearly cost the cat its life.

There are many sites that hold offensive materials that may be racist, anti-social or pornographic. Illegal terrorist organisations can publish material that incites violence – for instance it is possible to find a site that holds information on how to produce a bomb at home. The availability of usenet newsgroups that allow users to share information also poses a problem as many undesirable groups have been set up, such as violent political extremists, paedophiles and pornographers.

The need for some form of censorship on the web is constantly being debated. Many people feel that the Internet was set up to be a shared source of free information that should be available to everyone. For them, any form of censorship would erode this freedom. Not all governments are benign: censorship can allow governments to suppress legitimate criticism.

Others will argue that it is not acceptable that offensive material such as pornography, or articles that incite racial hatred, should be stored on the net.

The practical difficulties of imposing effective censorship are great due to the global nature of the Internet. There are different laws regulating the publication of material in different countries. Although it may be illegal to store certain material in one country, citizens can access such material from a site in another country where the material could have been stored legally.

The need to protect young people from viewing inappropriate material has resulted in the development of a range of software that prevents access to undesirable sites. This takes the form of an internet filter that stops a child downloading sites on the filter's 'bad list', for example those including violence, intolerance or sex. Some filters vet newsgroups, chat rooms and e-mails and stop children sending personal details such as addresses. Microsoft Explorer and Netscape Communicator have features in their browsers which allow parents to choose the sites they would like their child to access. By using such software censorship is taking place at the destination rather than at the source.

Case Study 4

Censorship and the Internet

This extended case study provides extracts from several articles in the *Guardian* concerning Internet censorship.

Yahoo! goes to battle over content freedom – June 2001

Internet portal Yahoo! won the first round of its long legal battle this week after a US judge agreed to rule whether foreign countries have jurisdiction over material published on American web sites. The case revolves around the portal's ability to host a site that auctions Nazi material online. Last year, a French court ruled that the company could not host a Nazi memorabilia site to which French citizens had access because the sale of Nazi material is illegal in France.

Yahoo! initially argued that as a US company, it was not subject to French laws and that it could not prevent French citizens logging onto its US site. The company's actions, it said, were perfectly legal in its home country. But Yahoo! executives bowed to European pressure after the French courts imposed daily fines of up to $13,000 on the US parent company.

Yahoo! then turned to a California district court for protection against the ruling. The company believes the French law contradicts the US right to free speech and that foreign courts should not have jurisdiction over the domestic actions of US companies.

The internet is a friend of freedom – January 2001

Six years ago Steven Gan, an investigative reporter for a Malaysian paper, wrote an exposé of the conditions at Semenyih immigration detention camp in Selangor describing the brutality and squalor and rampant sickness there. But Gan's story wasn't published. The Malaysian press is not free.

Steven Gan went on to found malayasiakini.com, a paperless newspaper. Gan and his team's strength is that they cover the things the country's print papers have to bury or set to one side. The newspaper survives through the unique strength of the Internet. A country like Malaysia, which has to be open for business, must also be open to the net.

5 years for man who lured girl via Internet – October 2000

A paedophile who used an internet chat room to entice a 13-year-old girl to his home for sex was yesterday jailed for five years in the first conviction of its kind. The man logged on to a chat room and exchanged messages with the girl for two months, gaining her trust before arranging a meeting and then abusing her. Passing sentence the judge described the man as a 'predator' and ordered him to remain on the sex offenders' register for life.

Web inventor denounces net censorship – October 2000

The British scientist who invented the world wide web has called for the abolition of censorship on-line. As parents' groups and politicians press for new ways to police web sites, Berners-Lee dismisses the recent outcry over paedophiles targeting youngsters in web chat rooms, child pornography and fraud, and rejects calls for a 'net regulator'.

'Regulation is censorship – one grown-up telling another what they can and cannot do or see.' Illegal material – child pornography, video nasties' – should remain illegal, but he insists 'the world is a diverse place and we should trust people, not try to police them... There are many cultures and they are continually changing. Two neighbours next door to each other might have very different ideas. So any attempt to make a global centralised standard is going to be unbelievably contentious. You can't do that.'

The Regulation of Investigatory Powers Act (RIP)

RIP sets out the government's plans to allow official surveillance of Internet traffic. The government state that RIP gives police and security services new powers to monitor and intercept e-mails and web sites, channels through which on-line criminals, terrorists, paedophiles and pornographers have traditionally been able to communicate. Some data on computers, such as child pornography, is often encrypted. The bill makes it illegal to refuse a police request to decrypt data.

RIP allows the government to monitor all UK internet traffic – sites downloaded, address books, e-mails, discussion groups and chat rooms – without a warrant.

○ Prepare a report on the issue of censorship on the Internet, putting forward arguments for and against.

Internet use in the workplace

There is a growing concern about the amount of work time that is wasted by some employees who are surfing the net, indulging in on-line shopping and sending and receiving an excessive number of personal e-mails during work time. Accessing inappropriate sites, for example those displaying pornographic images, is another worrying development.

In some organisations e-mail has become the new form of 'office chatting' and gossip. There is a growing trend of sending e-mails with joke attachments. As well as posing a security risk, as these attachments can carry viruses; such messaging can be very time wasting. In some cases offensive material is sent that can be seen as a form of harassment.

IT departments can track the sites that a user accesses as well as read the contents of e-mail. Some organisations are using these facilities to monitor staff usage. Four lawyers were recently suspended from a top law firm after they had been found to be circulating pornographic e-mails. Many people fear that such surveillance techniques will have a bad effect on employee morale and are an invasion of privacy.

Nearly 40 per cent of employees have e-mail addresses supplied by their employers. Many people suffer from e-mail overload at work, receiving junk mail (or SPAM) and circular letters as well as unnecessary messages from work colleagues. Increasingly, organisations are having to provide training in appropriate e-mail use.

Security

As mentioned earlier in this chapter and in Chapter 5, the use of the Internet provides a security problem as using it opens a computer system to attack from viruses, hacking and other fraudulent activities. The majority of viruses are passed via the Internet these days either directly through program or other files that are down-loaded or through documents that are attached to e-mail.

Very often an organisation will have access from a Local Area Network to the Internet so that all workstations have rapid access. This immediately puts all that is stored on the Local Area Network under threat as, once a connection is made, it can be accessed externally.

E-commerce, the use of the Internet to carry out commercial activities such as ordering products and services, depends in the main on the use of credit cards for payment. This has proved to be a barrier for many people: the use of e-commerce has not increased as quickly as was first forecast as customers are afraid that their credit card details will get into the wrong hands. They need to be reassured by a company that this will not happen.

Activity 4

Carry out a class debate on the topic of censorship and the Internet. A group of students should research the arguments in favour of censorship and another group those against. The arguments should be clearly made to the whole class.

What a company can do to reduce credit card fraud

- Only display part of credit card details and keep whole card details only in encrypted form.

- Use a firewall so that no hacker can gain access to an individual's credit card details when they are stored on the company computer.

- Provide regular customers with a password so that they do not have to enter their credit card details every time they make a purchase.

- Check the physical address to which goods are to be delivered against credit card or with the bank to ensure that the correct person is using the card.

Summary

The weak points within an IT system are associated with people, hardware and software. At risk are:

- automatic data entry

- user work stations

- networks

- IT personnel developing software

- use of laptop computers

- data stored off-line

The growth of the Internet raises a number of important issues that include:

- information stored on the Internet is not necessarily correct

- should there be any form of censorship of material that can be stored on the Internet?

- how can the Internet be made secure from hacking, particularly for e-commerce?

- is growth in access to the Internet resulting in inappropriate office behaviour?

Information systems, malpractice and crime questions

1. Describe, with reasons, three measures, other than passwords, that may be taken to maintain the integrity of data against malicious or accidental damage. *(6)*

   NEAB 2000 Paper 1

2. The illegal use of computer systems is sometimes known as computer-related crime.

   a) Give three distinct examples of computer-related crime. *(3)*

   b) Give three steps that can be taken to help prevent computer-related crime. *(3)*

   NEAB Specimen Paper 1

3. Describe the dangers of crime that the Internet presents. *(3)*

4. Explain, with reasons, two levels of access that could be given to different categories of users of an on-line stock control system. *(3)*

   AQA ICT Module 1 May 2001

5. A school wishes to allow its students unrestricted access to the Internet for research work during their lunchtimes. The headteacher is concerned that this might cause problems.

   State two problems that the headteacher might be concerned about, and for each one explain a measure that could be taken to prevent the problem. *(6)*

   AQA ICT Module 1 Jan 2001

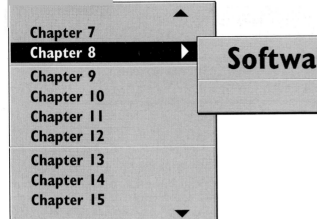
Software and data misuse

Computer Misuse Act 1990

The **Computer Misuse Act 1990** was introduced as a result of concerns about people misusing the data and programs held on computers. The act aims to protect computer users against malicious vandalism and information theft. Hacking and knowingly spreading viruses were made crimes under the act, which aims to secure computer material against unauthorised access and modification.

The act created three new offences:

- unauthorised access to computer material

- unauthorised access with intent to commit or facilitate commission of further offences

- unauthorised modification of computer material.

The penalty for each of the three categories is increasingly severe.

Unauthorised access to computer material

In this category, you are committing an offence if you try to access any program or data held in any computer without permission and you know at the time that this the case. The maximum penalty is six months in prison and a £5000 fine. This category applies to people who are 'just messing around', 'exploring the system' and have no intention of doing anything to the programs or data once they have gained access. It covers the act of guessing passwords to gain access to a system. An authorised user of a system may still be in breach of this category of the act if she accesses files in the system that have a higher level of access than she has been allocated rights to. A student gaining access to a fellow student's area, or breaking into the college administrative system, is breaking this category of the act.

Unauthorised access with intent to commit or facilitate commission of further offences

This category covers offenders who carry out unauthorised access with a more serious criminal intent. The access may be made with an intention to carry out fraud. For example, someone may break into a bank, personnel or medical system with the intention of finding out details about a person that could be used for blackmail. Persistent hacking is included in this second category. Prosecution under this category can lead to up to five years in prison.

Unauthorised modification of computer material

This third category concerns the alteration of data or programs within a computer system rather than simply viewing or using the data or program. In this case the program code could be actually changed. This could stop a program from running or to act in an unexpected manner. Alternatively, data could be changed:the balance of a bank account altered; details of driving offences deleted or an examination mark altered. This category includes the deliberate distribution of computer viruses.

Case Study 1

Computer Misuse Act Cases

In the summer of 1994 Stephen Fleming, a temporary employee at British Telecom, gained access to a computer database containing the telephone numbers and addresses of top secret government installations. Mr Fleming, who had worked at BT for two months, found passwords written down and left lying around offices and used them to call up information on a screen and copy it.

In November 1995 Christopher Pile, who called himself the Black Baron, became the first person convicted under the Computer Misuse Act. Pile created two viruses named *Pathogen* and *Queeg* after characters in the BBC sci-fi comedy Red Dwarf. The viruses wiped data from a computer's hard drive and left a Red Dwarf joke on screen which read 'Smoke me a kipper, I'll be back for breakfast ...unfortunately some of your data won't'.

Two 18 year-olds arrested in Wales were alleged to be computer hackers involved in a million-dollar global Internet fraud that involved hacking into businesses around the world and stealing credit card details. The youths were arrested under the Computer Misuse Act 1990; a home PC computer was supposedly used for the crime. Apparently they had accessed the credit card databases of nine e-commerce companies, and had published the details of thousands of credit card accounts on the Internet.

- For each of the three cases described, explain which category (or categories) of the Computer Misuse Act has been broken.

The Copyright, Designs and Patents Act 1988

Copyright laws have long protected the intellectual rights of authors, composers, artists, etc. They also apply to computer software. When you buy software you do not buy the program, only the right to use it under the terms of the licence. It is illegal to copy or use software without having obtained the appropriate licence. Criminal penalties can include unlimited fines and 2 years' imprisonment or both.

A software licencing agreement is a legal contract between the software producer and the user that sets out how the piece of software may be used.

Figure 8.1 Software licensing agreement

Software licensing methods

Computer owners, particularly if they own more than one, for example, a company with a network must be aware of the terms of the licence agreement which comes with the software.

A number of different licence agreements are normally available:

A **single-user licence** allows a copy of the software to be installed on a single machine. The software is usually supplied as a package consisting of the programs on disks (floppy or CD) together with user manuals. If the user wishes to install the software on an extra machine he must purchase another copy.

If it specifies that only one copy may be in use at one time, then the user is entitled to install the software on two computers – perhaps his desktop PC and a laptop – as long as at any given time the software package is only being used on one of the computers.

A **multi-user licence** allows an organisation to install the software package on an agreed number of computers. This would normally cost less than purchasing the same number of single-user licences.

A **site licence** allows a user to purchase a single copy of the software with permission to install the software on all the computers at a single location. This is a cheaper method than purchasing a single copy for each machine. Site licensing is common in the education sector.

A method of licensing that is increasingly being used with networked systems is the **licence by use**. This allows the software to be installed on a large number of stand-alone computers. However, only an agreed number of users are allowed to run the software at any given time. A 10-user licence of this type would allow the software to be installed on 50 computers as long as there were never more than 10 people using the package at once.

An alternative licensing method for an organisation running a local area network is a **network licence**. One copy of a software package will be stored on the file server. If a 10-user network version of a package is bought, then only 10 people can use the package at any one time. There can be hundreds of computers on the network and the package will be available to any one of the computers. When an eleventh user tries to use the software package, access will be denied.

Other forms of network licence allow the software to be used simultaneously at any number of computers.

When you buy software:

- you do not normally have the right to give a copy to a friend
- it is normally illegal to make a copy of the CD with a CD writer and then sell it
- it is illegal to use the software on a network unless the licence allows it
- you cannot rent out the software without the permission of the copyright holder
- make sure that you get a copy of the licence: a licence is a valuable document
- always make sure you buy software from reputable dealers.

Some software is made available without the restrictions of copyright. Such software is known as **public domain software**. These programs can be distributed, copied and used free of charge. Any users can alter the program code of such software. Most **shareware** is obtained by down loading from the Internet. Shareware is software that is licensed to

be used without charge for a trial period. After that time a fee must be paid if the user wishes to continue using the software.

Freeware is software that is available under similar rules but, unlike public domain software, the code is not allowed to be altered. The program is copyrighted, but the author does not charge for it, allowing a user to copy and use it at will.

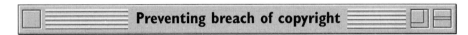
Software piracy

Software piracy is the unauthorised copying of software. It applies whether the copying is carried out on a large scale, where copies are sold for financial gain, or by an individual user for personal use. A user is allowed to make copies of software for back-up use only.

Software piracy is a very large problem: many people do not seem to realise that whenever they take a copy of software from a friend to use on their own computer they are breaking the law.

Preventing breach of copyright

Software companies are getting more and more sophisticated in their attempts to stop the software pirates.

- The licence agreement is clearly printed on the packaging.

- Some games are designed only to run if a code is typed in. This code changes each time the program is run and can be found by looking in the manual or using a special code-wheel which comes with the software.

- Some programs will only run if the CD-ROM is in the CD drive.

- Some programs will only run if a special piece of hardware called a **dongle** is plugged into the back of the computer.

FAST

Figure 8.2 FAST logo

The Federation Against Software Theft (FAST) was founded in 1984 by the software industry. It is supported by over 1,200 companies. It is a not-for-profit organisation which aims to prevent pirate software and has a policy of prosecuting anyone found to be breaching copyright law.

FAST also works to educate the public about good software practice and legal requirements. FAST identifies the following types of software piracy:

Professional counterfeits

These are professionally made copies of software including the media, packages, licences and even security holograms. They are made to resemble the genuine article as far as is possible.

Recordable CD-ROMs

Pirates compile large amounts of software onto one recordable CD-ROM and make multiple copies of the CD-ROM. There is usually no attempt to pass these off as the genuine article. These recordable CD-ROMs will then be sold to customers in different ways through mail order, 'under-the-counter' at retail outlets or at markets, and now are even advertised on the Internet.

Hard disk loaders

These are dealers or retail outlets who load infringing versions of software onto a computer system to encourage customers into buying their computer hardware. Customers will not have the appropriate licences, will not be entitled to technical support or upgrades, and may find the software on the computer to be incomplete or to contain viruses.

Internet piracy

This is the down-loading or distribution of infringing software on the Internet. Just because software is present at certain sites on the Internet does not necessarily mean it is free or legal for you to down load. It may have been placed on the Internet without the copyright owners' consent. Software pirates also advertise their services on the Internet by offering cheap mail-order software.

Corporate over-use

This is the installation of software packages on more machines than there are licences for. For example, if a company purchases five single-user licences of a software program but installs the software on ten machines, then they will be using five infringing copies. Similarly, if a company is running a large network and more users have access to a software program than the company has licences for, there will be corporate over-use.

Activity I

Find out more about the fight against illegal software at FAST website http://www.fast.org.uk/. Outline the kinds of activities that FAST is involved in.

Case Study 2

Software Pirate Jailed for 2 years

Computing, 14 July 1999

Software pirate Terrence Brown has been jailed for 2 years on 9 counts of Copyright and Trademark offences, as a direct result of investigations by FAST (The Federation Against Software Theft) and Kingston Trading Standards Officers.

Brown first came to FAST's attention in November 1997, when he placed an advert in a trade magazine offering computer software at very low prices under the bogus company name, 'CD Direct'.

When officers visited Brown's address, they found two CD writers and over 500 pirated CDs containing pirate software. While on bail Brown had started counterfeiting again and a further two CD writers, scanners and more CDs were seized. On 5 July 1999 Brown was sentenced to a total of 2 years in prison.

Activity 2

Produce a poster for your classroom, a slide presentation or a series of web pages aimed at an audience of students in your school or college. Your poster should make clear to them which of the activities a student might carry out are illegal under the Computer Misuse Act or Copyright Acts.

Summary

The Computer Misuse Act of 1990 makes the following illegal:

- accessing computer material without permission e.g. hacking

- unauthorised access to a computer to commit another crime

- editing computer data without permission e.g. spreading a virus

The Copyright, Designs and Patents Act 1988 requires all users of software to have a valid licence.

Computer owners, particularly if they own more than one, for example, a company with a network must be aware of the terms of the licence agreement which comes with the software. A number of different licence agreements are normally available:

- single user licence where the software can only to be used on one machine

- network licence may be for up to 15 or 20 stations on a network or every station on the network, depending on the licence – this licence is obviously much more expensive than a single user licence

- site licence, which enables the software to be used on any computer on the site

Software and data misuse questions

1. The Computer Misuse Act defines three types of offence. With the aid of examples, describe each of these three types of offence. (9)

AQA ICT Specimen Paper 1

2. State the three levels of offence under the Computer Misuse Act of 1990. Illustrate each answer with a relevant example. (6)

NEAB 1999 Paper 1

3. Explain some of the things software companies have done to prevent illegal copying of their software. (4)

4. a) Describe what is meant by a software licensing agreement. (2)

b) Mr Patel has a single user software licensing agreement for a word processing package which specifies that there must only be one copy in use at any one time. Would he be guilty of breaking the agreement if he installed the package on his laptop PC, as well as on his stand-alone PC at work. Explain your answer. (2)

c) A college network has a server and 20 stations. What type of licensing agreement would be suitable for a word processing package that may be used at the same time on all 20 stations? (2)

d) Breaking the licensing agreement is one type of offence that a computer user can commit. Some other offences are covered by the Computer Misuse Act. Explain, using examples, level 1, 2 and 3 offences under the Computer Misuse Act. (9)

AQA ICT Module 1 May 2001

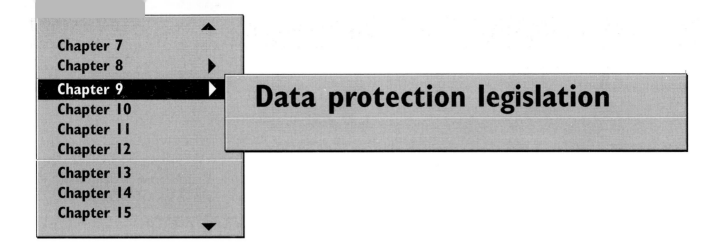

Data protection legislation

'Think before you give away personal information, you never know where it will end up!'

(DATA PROTECTION ADVERTISING CAMPAIGN SLOGAN)

Why data protection laws were introduced

The development of new technology has meant that many organisations store and process personal details on computer. This may be data on customers, employees, suppliers, competitors, etc. This has worried many people whose main concerns are:

- Who will be able to access this data? Will information about me be available remotely over a network and therefore vulnerable to being accessed by hackers for example? Can my school records be examined by a potential employer? Can they be sold on to someone else?

- Is the data accurate? If it is stored, processed and transmitted by computer, who will check that it is accurate? People often think it must be true if 'it says so on the computer'.

- Will the data be sold on to another company? For example, could my health records be sold to a company where I have applied for a job? Could my personal details, collected by my employer, be used by a commercial company for targeting junk mail?

- As it is very easy to store vast amounts of data, will data about me be stored even if it is not needed, for example if I apply for a job but don't get it?

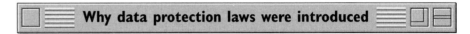

What is personal data?

Personal data covers both facts and opinions about an individual. Facts would include name, address, date of birth, marital status or current bank balance. Results in examinations, details of driving offences, record of medications prescribed and financial credit rating are further examples of facts that could relate to an individual. Personal opinions such as political or religious views are also deemed to be personal data.

Data Protection Acts 1984 and 1998

It was concerns about personal data, which grew as the use of computers increased in all areas of life, that led to the Data Protection Act 1984. The 1984 Act set out regulations for storing personal data that was automatically processed. It made no provision for data that was stored on paper. However, as the use of ICT for storing personal data continued to expand, it was felt that the 1984 act did not go far enough in protecting the rights of individuals.

The 1998 Data Protection Act extended the 1984 Act and enshrined the European Union on Data Protection into UK Law. This meant that UK law was in line with other Data Protection laws in other countries in the European Union. The new act extended the scope to the protection of paper-based data. The act also increased penalties for breaking the law.

What the Data Protection Act 1998 says

The Data Protection Act 1998 mostly came into force on 1 March 2000 and sets rules for processing personal information. It applies to some paper records as well as all those held on computer and works in two ways:

- it gives individuals certain rights;
- it stipulates that those who record and use personal information must be open about how the information is used and must follow good information handling practices.

It is a legal requirement laid out in the Data Protection Act that an organisation which keeps data relating to individuals must apply for entry on the Data Protection register.

The law refers to

- data controllers (formerly called data users) – people or organisations who decide how and why personal data, (information about identifiable, living individuals) are processed
- data subjects – people whose personal data is being processed.

Data controllers must register with the Information Commissioner (formerly the Data Protection Commissioner). The following must be registered:

- the data controller's name and address
- a description of the data being processed
- the purpose for which the information will be used
- from whom the information was obtained
- to whom the information will be disclosed and countries where the data may be transferred.

Data controllers must follow the eight principles of the Act. Unlike the 1984 Act, the 1998 Act applies to paper records as well as electronic records, but only from 23 October 2007.

The eight Data Protection principles

1. Personal data must be processed fairly and lawfully. A data subject must be informed that data is being collected and what it is to be used for. A data subject will usually have to have given written permission before sensitive personal data, such race, sexual orientation or political beliefs can be gathered and processed.

2. Personal data can be held only for specified and lawful purposes, for example data on new-born babies that is held by a maternity unit should not be used to generate mail shots advertising baby products.

3. Personal data should be adequate, relevant and not excessive for the required purpose. An organisation's employee records are unlikely to require the marital status and details of an employee's children.

4. Personal data must be accurate and up to date.

5. Information should not be kept for longer than is necessary for the purpose for which it is collected. There is no need for a personnel department to keep the application forms of unsuccessful job candidates.

6. Data must be processed in accordance with the rights of the data subjects (see rights of individuals page 98.) This gives individuals the right to inspect the information held on them. They have the right to require that inaccurate data must be corrected. It is not unusual for an individual to be refused a credit card because some incorrect data that indicates that they are not a good credit risk is stored.

7. Appropriate security measures must be taken against unauthorised access. This means information has to be kept safe from hackers and employees who don't have rights to see it. Data must be also safeguarded against accidental loss.

8. Personal data cannot be transferred to countries outside the European Union unless the country provides an adequate level of protection. This was a new principle in the 1998 Act.

Rights of individuals

Under the sixth Data Protection principle, data subjects have the right to

1. see data held on themselves within forty days for payment of a small fee

2. have any errors in the data corrected

3. compensation for any distress caused if the act has been broken

4. prevent processing likely to cause damage or distress

5. prevent processing for direct marketing by writing to the data controller to inform them that advertising material is not to be sent

6. prevent processing for automated decision taking by writing to the data controller to inform them that no decisions should be taken based on automatic processing. Some banks determine whether or not a customer should be given a mortgage on the basis of a computer program. The data subject has the right to prevent this happening.

Is your consent needed to process your data?

Suppose you apply for a supermarket loyalty card. The supermarket will need to store and process your personal details as part of their normal work. They do not need your consent to do this as you have agreed to this when you applied for the card.

However, the supermarket cannot pass your details on to another company without your consent. In practice this means that when you fill in the application form you can tick a box to prevent your personal details being sold on to another company, for example for direct marketing.

Note:

1. If you cancel the card, your details are no longer needed and should be deleted by the supermarket.

2. At any time you can write to an organisation that is sending you junk mail asking them to stop processing any data about you.

Exemptions from the Act

The following exemptions exist from the principles of the Data Protection Act.

- if the information is used to safeguard national security

- if the information is used for the prevention and detection of crime

- if the information is used for the collection of taxes

- if the information is used in journalism for historical and statistical purposes

- Personal data relating to someone's family or household affairs does not need to be registered.

The 1998 Act does not exempt payroll and pensions data, members clubs and mailing lists, which were exempt under the 1984 Act.

⟩ What is the role of the Information Commissioner? ⟩

The Commissioner has the responsibility of ensuring that the Data Protection legislation is enforced.

The Information Commissioner keeps a public register of data controllers. Each register entry must include the name and address of the data controller as well as a description of the processing of personal data carried out under the control of the data controller. An individual can consult the register to find out what processing of personal data is being carried out by a particular data controller.

The Data Protection Act 1998 requires every data controller who is processing personal data to notify the commission unless one of the exemptions listed in the Act applies. At the commission office a complete copy of the public register is kept and it is updated weekly.

Other duties of the Commissioner include promoting good information handling. As well as keeping the register of data controllers, the Information Commissioner also gives advice on data protection issues, promotes good information handling practice and encourages data controllers to develop suitable codes of practice. The Commissioner also acts as an Ombudsman.

Note: The Information Commissioner was previously called the Data Protection Registrar under the 1984 Act.

⟩ A company view of the Data Protection Act ⟩

When a company needs to apply for entry in the Data Protection Register it must supply information on:

- The purpose for which the data is to be stored

- What data fields are to be stored

- Who will have access to the data

- The name of any organisation(s) to whom the data will be passed on

- The sources of the data and how it has been acquired.

The organisation will need to provide a code of practice for its employees to ensure that the DPA is complied with. The code of practice will lay down how data is to be collected, processed and kept secure. It should state how long each category of data should be kept.

Considerable extra work has been created for organisations if they are to comply with the DPA.

> ## Implementing the DPA in a college

At a college, data relating to staff members and students is kept. Staff data is collected during recruitment and all application forms contain a signature box where the applicant approves that the data should be stored and processed. A table of all the data items stored is shared with staff; an extract is shown below:

Data	Reason for collection	Authorised Third Party Access	Reason for Third Party Access
Address	Identification Contact	Payroll Pensions Agencies	Correspondence Records
Qualifications	Employment	Learning and Skills Council (LSC)	Statistics
Medical History	Employment Adjustments	LSC	Statistics
Current medical Condition	Employment Absence monitoring	Payroll	Statutory sick pay
Car details	Parking		
Date/ Place of Birth	Eligibility for employment		
Bank / BS	Payment	Payroll	Payment

Paperwork on unsuccessful job applicants is retained for 12 months; staff personal records are kept for six years after the date of leaving; staff pay records (including tax and National Insurance) are kept for six years.

The personnel department has to be sure that all electronic data is deleted and paper records shredded after the stated times have elapsed.

Once a year a printout is produced of the standard data held for each member of staff and issued to them to be checked for accuracy.

Activity 1

Obtain a copy of your school or college's Code of Practice for Data Handling.

○ What are the main headings in the Code of Practice?

○ Produce a table, similar to the one stored above, with data items that are stored about students.

○ Who is the data controller for your school or college?

Case Study 1

Legal or illegal Use of Data

A company registers the following use with the Information Commissioner:

Purpose: The administration of prospective, current and past employees.

Typical activities are: Payment of wages, salaries, pensions and other benefits; training, assessment and career planning.

Type of information: Names, addresses, salary details and other information related to their work

1. The company wants to sell the names and addresses of their employees to another company for direct marketing. This is **illegal** as it is not the registered purpose so breaches Principle 3.

2. The police want the company to give them information about an employee in relation to a possible fraud. This would be **legal** as long as the police provided a certificate of exemption that this was connected with detection of crime.

You can find out more about the Data Protection Act and the Information Commissioner at
http://www.dataprotection.gov.uk/ .

Explore the site and answer the following questions:

◦   How can you find out what credit reference agencies are storing about you and how can you correct any mistakes?

◦   What is the name of the current Information Commissioner?

◦   List all the fields that are on the Notification Form that has to be filled in by a data controller

◦   What is the 'Information Padlock'?

Case Study 2

Lack of Security

Danny Hughes is an A Level student who had a holiday job on the production line at Betta Biscuit plc. One lunchtime Danny decided to explore the factory and found his way into the computer room. There was no one about. Danny sat down at a terminal and typed in a few usernames with no luck. Then he noticed a birthday card on the desk – to Bob with love from Jane. Danny typed in the username BOB and was asked for a password. He typed in JANE and to his surprise it was accepted!

A menu appeared on the screen. Danny chose payroll. He could load up the payroll information of all the employees. Danny loaded the file of his friend Chris and cut his hourly pay by half. Two workers came in and saw Danny, but no one said anything. Danny logged off quietly and slipped out of the room undetected.

◦   Suggest as many steps as you can that Betta Biscuit plc should take to improve their security.

◦   Who has broken the law – Danny or Betta Biscuit or both of them?

Case Study 3

Organ Group President is Fined for List on Computer

Daily Telegraph

The honorary president of the Association of Organ Enthusiasts was fined £50 and ordered to pay £683 costs yesterday for keeping his membership list on his home computer.

The man told magistrates he did not realise that he had to register and pay a £75 fee to the Data Protection Registry. Admitting two breaches of the Data Protection Act, he told the court that he had spent hours keying in details of organ enthusiasts who subscribe to his bi-monthly magazine.

He said: 'I didn't think names and addresses were sensitive information. I am sorry for what I have done'.

● Use the Internet to search for details of other breaches of the Data Protection Act. Use sites such as www.guardian.co.uk/ or www.independent.co.uk

Summary

● The Data Protection Act concerns the storage of personal data. Data controllers must:

- register with the Information Commission (formerly the Data Protection Commissioner).

- follow the principles of the Act

● the principles of the Act say that personal data must be accurate and up-to-date and only used for the registered purpose

● data subjects have various rights under the Act including the right to inspect data about themselves, have any errors corrected and claim compensation for any distress

● personal information involving national security matters, the detection of crime and the collection of taxes is exempt from the Data Protection Act

Data protection legislation questions

1. (a) State the eight principles of the 1998 Data Protection Act. *(8)*

(b) Describe two exemptions to the Act. *(4)*

(c) Describe two obligations of data users under the legislation. *(4)*

AQA IT 2000 Paper 1

2. A college maintains an extensive database of its full-time students. The database contains personal data, the courses students attend, and higher education or employment applications.

a) Describe how the college might keep the personal data of the students up to date. *(3)*

b) The college wishes to sell the personal data to a local sports retailer. An agreement is to be written between the college and the retailer. Describe three issues, relating to the data that should be included in this agreement. *(3)*

NEAB 1997 Paper 1

3. a) State five of the principles of the 1998 Data Protection Act. *(5)*

b) Describe two exemptions to the 1998 Data Protection Act. *(4)*

AQA ICT Module 1 January 2001

4. a) State why an organisation must apply for entry on to the Data Protection Register. *(1)*

(b) State three items of information that must be provided by the data user about the data that is to be stored. *(3)*

AQA ICT Module 1 May 2001

5. A national distribution company advertises its products by sending personalised letters to thousands of people across the country each year. This type of letter is often known as 'junk mail'. The distribution company purchases the list of names and addresses from an agency.

a) Describe two ways in which the use of information technology has increased the use of 'junk mail'. *(4)*

b) The company wishes to target letters to people who are likely to buy its products. How might this be done? *(2)*

c) A person receiving this type of mail writes to the company to complain that it is acting illegally under the terms of the Data Protection Act. Give three statements the company may use in its reply to show that it is operating within the terms of the Act. *(3)*

NEAB 1996 Paper 1

Hint: An example of a statement may be: We follow the fifth principle of the Data Protection Act by deleting all personal information that we no longer use.

6. Charlie Reid wishes to buy a washing machine on hire purchase. He is refused credit as he has a bad credit rating on the computer. He asks why and he is told that his son, who lives at home, has a bad credit rating. When he asks why his son has a bad credit rating, he is told that his son used to live at an address 100 miles away in Lancashire. Another resident at the Lancashire address owes his bank £1000. He complains to the bank that they've broken the Data Protection Act. They deny that they have. Explain whether you think the Data Protection Act has been broken or not.

7. In 1997 the Information Commissioner, Elizabeth France, threatened to take two utility companies to court for breaking restrictions in the Data Protection Act. The companies were sending direct mail leaflets and advertisements to their customers. The official notice from the registrar to the two companies said they should stop using the customer list for advertising. Write briefly to one of the companies explaining which principle of the Data Protection Act may have been broken.

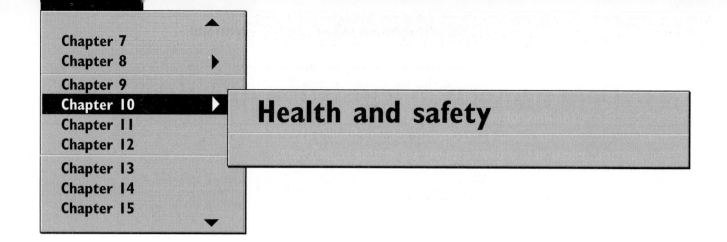

Health and safety

Many computer users have blamed computers for various problems with their health. Many of these problems are avoidable. The Health and Safety (Display Screen Equipment) Regulations of 1992 made it a legal requirement for employers to take various measures to protect the health of workers using computers.

Figure 10.1 What is wrong?

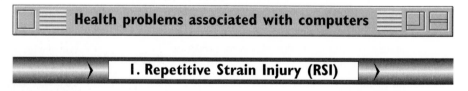

Health problems associated with computers

> **1. Repetitive Strain Injury (RSI)** >

It is widely accepted that prolonged work on a computer can cause repetitive strain injury. Using a keyboard which is positioned so that hands have to be held at an awkward angle can cause this injury. RSI affects the shoulders, fingers and particularly the wrists of those typing all the working day. The symptoms are stiffness, pain and swelling.

The risk of RSI can be reduced by using special 'ergonomic' keyboards and using wrist supports while typing.

Desks should have sufficient space on them to allow the user to rest his wrists when not typing.

Managers should ensure that work is varied during the day so that no user spends all the working day at the keyboard.

Take regular breaks from computers.

2. Eye strain

Looking at a screen for a long time can lead to eye strain, particularly if the screen is of a poor quality and flickers or has the wrong contrast setting. Problems also occur if the lighting in the room is at the wrong level, or the screen is poorly sited causing glare or reflection on the screen. The symptoms are headaches and sore eyes.

The risk of eye strain can be reduced by having suitable lighting, using non-flickering screens and fitting screen filters to prevent glare and reflection.

Appropriate blinds could be fitted to windows to reduce glare. Appropriate spectacles should be worn when using a computer.

The eye-line should be approximately level with the top of the screen and the screen slightly tilted.

Employers are obliged to pay for any employee who works long hours in front of a computer screen to have a free eye test.

Take regular breaks from computers.

3. Back problems

Sitting in an uncomfortable position at a computer or bad posture can lead to serious back problems. This is likely to be the case if chairs are at the incorrect height. The symptoms are back pain or stiffness, possibly a stiff neck and shoulders and sore ankles.

The risk of back problems can be reduced by having an ergonomically designed, adjustable chair that supports the lower back, and can be to the right height. Screens that tilt and turn can be adjusted to the correct position. Document holders next to the screen can reduce neck movement that leads to problems.

Take regular breaks from the computer

The use of a footrest can help prevent back damage.

4. Ozone irritation

Laser printers emit ozone which can act as an irritant. Bubble-jet or dot-matrix printers do not emit ozone. The symptoms are problems with breathing, headaches and nausea.

The risk of irritation can be reduced by locating laser printers at least one metre away from where someone is sitting to avoid the ozone emissions.

Good ventilation is essential.

5. Radiation

In the USA research has revealed that an unusually high percentage of pregnant computer users have abnormal pregnancies or have suffered miscarriages. This may be pure coincidence but it has been suggested that it is caused by electromagnetic radiation from monitors.

Low emission monitors give off less radiation.

Screen filters can also cut down radiation.

6. Epilepsy

It has long been suggested that flickering computer screens have contributed to the incidence of epilepsy in users.

Again the use of low emission monitors and screen filters are likely to reduce the risk of epilepsy in users.

7. Stress

Using computers can be frustrating and repeated frustration can lead to stress. Frustration can occur if

- a user has inadequate training in the use of a piece of software
- the response time is very slow
- the human computer interface is inappropriate – perhaps too cluttered
- there are too many stages or keystrokes required to carry out a simple task

- a user is constantly losing documents
- hardware failure is a common occurrence – for example the printer gets jammed or the computer crashes several times a day.

Remove frustration by providing users with adequate support and training, choose well designed software that has an interface that is appropriate for the user and install hardware that is capable of meeting the demands of the tasks.

How big is the problem?

As computer use has increased the incidence of RSI has risen. In 1995 the US Occupational Safety and Health Administration said that of all the illnesses reported to them, 56 per cent were RSI cases, compared with 28 per cent in 1984 and 18 per cent in 1981. Among the generally accepted causes of these cases were poorly designed workstations, ill-fitting chairs, stressful conditions and extended hours of typing.

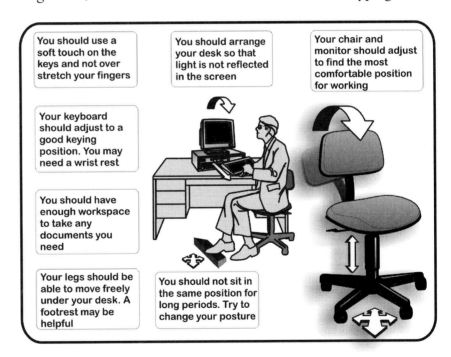

Figure 10.2 Adjustable chairs can prevent injuries

What the law says

In 1992 the European Union introduced regulations to try to prevent health problems associated with computer equipment.

The regulations were called the Health and Safety (Display Screen Equipment) Regulations 1992.

Activity 1

Find out more about health hazards and ICT at the Health and Safety Executive's web site: http://www.hse.gov.uk/

- What services does the site provide for a) an employer b) an employee?

The Trade Union Council (TUC) health and safety newsletter site, http://www.hazards.org/ is worth exploring too.

Employers must:

1. Provide adjustable chairs
2. Provide tiltable screens
3. Provide anti-glare screen filters
4. Ensure workstations are not cramped
5. Ensure room lighting is suitable
6. Plan work at a computer so that there are breaks
7. Provide information on health hazards and training for computer users;
8. Pay for appropriate eye tests for computer users
9. Provide special spectacles if they are needed and normal ones cannot be used.

Note: These regulations apply only to offices and not to students in schools or colleges.

Designing software to minimise hazards

Software should be designed to reduce health risks. Software designers must be aware that bright colours can cause eye strain, annoying use of sound can lead to headaches and the use of flashing images (e.g. in games) can lead to epileptic attacks.

Designers must ensure that:

- screen layouts that are clear so that eye strain is minimised
- colour schemes are not too bright and have good contrast otherwise eye strain could occur
- text is a suitable size, minimising the chance of eye strain
- sufficient instructions and help facilities to enable learning are included otherwise the user could feel stressed
- menu systems are well structured otherwise lengthy navigation could be stressful
- annoying sounds and flashing images are used sparingly
- software is compatible with other packages otherwise slow use could lead to stress
- new versions of software should be backward compatible otherwise work done on the older version will have to be redone on the new which could cause stress.

See Chapter 21 Human/Computer Interface for a detailed discussion of this topic.

It is always difficult to develop regulations to cover an area that is evolving so quickly that by the time you have researched the problems and made recommendations on a given technology, it may be out of date.

Activity 2

Produce a poster or slide show that will provide an employee with the information they need to protect themselves from one of the following hazards:

- Eye strain
- RSI
- Back strain.

Case Study 1

Younger Staff More at Risk of Repetitive Strain Injury

The Guardian, March 3 2001

Young workers are more at risk from Repetitive Strain Injury than their older colleagues, according to figures released this week by the TUC.

Having to carry out repetitive tasks at speed, needing to use a good deal of force when working, not being able to choose or change the order of monotonous tasks and having to work in awkward positions are the key factors which health and safety experts agree make workers especially vulnerable to RSI.

The TUC data shows that although 65% of UK workers of all ages have jobs which involve a repetition of the same sequence of movements, a staggering 78% of younger workers do so and more than half the UK's 4m workers aged 16-24 are forced to work in awkward or tiring positions.

Case Study 2

The Inside Track: Industrial injuries

The Guardian September 17, 2001

Linda Newsome may be about to raise the stakes dramatically in compensation claims for office injuries. A tribunal has found in her favour in her dispute with her former employer, Sunderland City Council, and her lawyers have suggested compensation of £250,000. Whether the tribunal agrees this figure remains to be seen.

Although over the past decade banks and local authorities have had to pay damages to groups of workers, what is exceptional about Newsome's case, apart from the large sum mooted, is that she has brought the claim as an individual.

Newsome is suffering from RSI (Repetitive Strain Injury), often thought of as the secretaries' disease, because the best known form is tenosynovitis, an inflammation of the tendon sheaths of the wrist, caused by, among another things, excessive use of keyboards. In fact, secretaries type less and less these days. Tenosynovitis is more commonly found among data-inputters and journalists. Newsome is an accountant.

Tenosynovitis is the best known form of RSI because it is a glaring medical condition and was the first type to be recognised by the courts. Other forms of RSI can be more subtle.

Newsome's illness is a case in point. It began when her employer computerised the office in 1993.

Newsome was issued with a chair that was so high her feet didn't reach the floor. She got into the habit of leaning forward to compensate, which aggravated a latent back complaint. When she started to take time off sick, she was expected to clear the backlog on her return, which meant more time in her chair, and so on. Conditions were aggravated by cramped working conditions and the stress of having to report to two different managers. She was refused a new chair.

Newsome won her case on the grounds that she had been discriminated against as a disabled person. Last year the ceiling on compensation for such claims was removed, hence the huge potential damages. There is more to office furniture than office politics.

Summary

Regular use of IT equipment over a long period of time may lead to health problems, particularly:

- RSI (Repetitive Strain Injury) mainly in the wrists

- eye strain

- back problems

- irritation due to ozone

These problems can be reduced by taking sensible precautions, not using equipment for too long and introducing adjustable chairs and screens, wrist supports and screen filters.

New regulations have been introduced throughout the European Union setting standards for using IT equipment in offices. Employers have duties to:

- provide adjustable chairs

- provide adjustable screens

- provide anti-glare screen filters

- pay for appropriate eye and eyesight tests

- plan work so that there are breaks from the screen

- provide information and training for IT users

Display screen equipment users are entitled to regular checks by an optician or doctor, and to special spectacles if they are needed and normal ones cannot be used. It is the employer's responsibility to provide tests, and special spectacles if needed.

Health and safety questions

1. Describe three features of poorly designed software that can cause stress or other health problems to a user. *(6)*

   AQA ICT Module 1 January 2001

2. Poorly designed computer workstations can lead to health problems. State three features of a well-designed workstation, and for each one state the health hazard that will be prevented. *(6)*

   AQA ICT Module 1 May 2001

3. State three factors that could give rise to health and safety problems associated with the use of IT equipment, and for each one state the associated health and safety problem.*(6)*

   AQA ICT Module 1 2000

4. What is meant by RSI? List the parts of the body it affects, and describe the symptoms.

5. Organisations are legally obliged to provide a safe and healthy working environment for their employees. Poorly designed computer workstations can lead to health problems.

   State four features of a well-designed workstation, and for each feature state the health problem for the employee that will be prevented. *(8)*

   AQA ICT Module 1 2001

6. List six factors which could give rise to health or safety problems associated with the use of information technology equipment. *(6)*

   NEAB 1996 Paper 1

7. A trades union is concerned about its members who use computers all day. Produce a leaflet describing the dangers and the rights of computer users.

8. Trade Unions did not think the health and safety regulations were strong enough. The employers complained that the new regulations would cost £42 per worker. Imagine you are either

   a) the boss of a company using computers or

   b) a trades unionist using a computer at work.

   Write a letter to Sir John Cullen of the Health and Safety Commission telling him whether you think the new regulations are a good idea and whether they go far enough or too far, explaining why.

9. Describe three health hazards associated with computer use. *(6)*

   NEAB Specimen Paper 1

10. The introduction of computer terminals and personal computers has been associated with a number of physical health hazards.

    a) State three health hazards which have been associated with prolonged use of computers. *(3)*

    b) Describe five preventative actions which may be taken to avoid computer related health hazards, explaining clearly how each action will assist in preventing one or more of the hazards you have described in part (a). *(10)*

    NEAB 1998 Paper 1

11. The use of Information Technology equipment has brought Health and Safety risks for employees. Describe four such risks, and the measures that an employer should take to protect their staff from them. *(12)*

    NEAB 1999 Paper 1

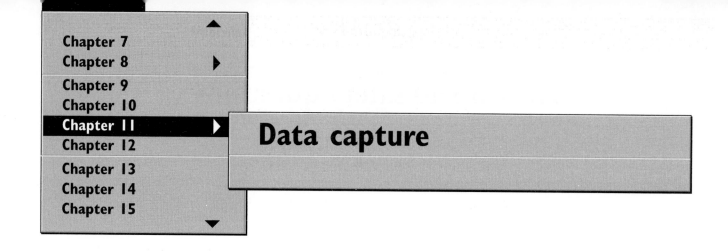

Data capture

Data capture is the practice of collecting and inputting data for computer processing. The data is converted into machine-readable digital format. This may be done automatically, as in scanning a barcode on a library book, or manually, as in a new student at a college filling in an application form that is then used as a source document for entering the data at a keyboard. Automated data capture removes the need to key in data and allows large quantities of data to be read quickly. It has an additional advantage of reducing the occurrence of errors as humans are prone to error (see Chapter 12). A range of data capture methods have been developed, each suited to a different range of applications and circumstances.

Factors influencing the choice of data capture method

Volume of data	If the volume of data is large, then specialised data capture equipment could be appropriate. The overheads of installation and maintenance costs as well as staff training can be justified. With a high volume of data some methods would not be able to cope within a reasonable time.
Speed	Speed can be an important factor in data entry. A flat bed barcode scanner can be used at higher speeds at a busy supermarket checkout that a hand-held scanner. Off-line data entry methods, such as OCR and OMR, can produce very fast data entry.
Nature of system	A particular system may have an obvious data capture method associated with it. For example, the need for security against fraud means that MICR is the chosen method for entering data from cheques. In a dirty or dangerous environment such as a factory, the use of a keyboard may not be appropriate. Stock and goods may be identified by barcode. The use of magnetic media may be impractical due to magnetic fields.
Ease of use	The conditions under which data is to be entered, together with the range of skills of the users may influence the choice of data capture method.
Technological development	The choice of data capture method for a particular system may have been different a few years ago and may be different again in a few years' time. New methods are being developed all the time. However, it is important that methods are reliable and have the confidence of the users.
Cost	The cost of installing a particular method of data capture will be a major factor in any decision. Cost could relate to staffing, or hardware, as well as the cost of changing from the old system to the new. The greater the volume of data the more likely that a higher cost method can be justified.

The volume of data to be captured will play a major part in determining the method used. A flat bed laser barcode scanner would be a sensible choice for entering the details of products sold in a large, busy supermarket but would not be a sensible choice for an exclusive boutique. The method of capture chosen will also be influenced by other factors such as speed and accuracy. The table on page 112 outlines the factors to be considered when choosing an appropriate method for a specific system.

The purpose of all methods of data entry into a computer system is to convert the data from the form that it exists in the world – for example, letters on a page, pencil marks on a document, black and white stripes in a barcode or sound waves – into a computer readable form. As all data held within the computer has to be stored in binary (made up of 0s and 1s) the devices described below all convert the real world data into binary codes which can then be processed by the computer software. This process is known as the **encoding of data** (see Chapter 1).

Optical Mark Recognition (OMR)

Optical Mark Recognition is a form of mark sensing where pre-printed documents are used. These documents contain boxes where marks can be made to indicate choices or a series of letters or numbers that can be selected to indicate choices. The reading device detects the written marks on the page by shining a light and recording the amount of light detected from different parts of the form. The dark marks reflect less light. The OMR reader transmits the data about each of the pre-programmed areas of the form where marks can be expected. The forms are printed with a light ink which is not detected.

Multiple choice exam papers use a special OMR answer sheet. Each student is given a mark sheet and they put a mark with a pencil through their choice for each answer (usually represented by a letter) on a grid. All answer sheets are gathered together and read by an OMR reader. Appropriate software is used to check the marks against the expected ones for the correct answers and the students' scores can be calculated.

The National Lottery and football pools coupons use a similar system. In some hospitals patients select from the day's menu on OMR cards that are read into a computer that is programmed to produce accurate information quickly and easily to be used by the catering services. OMR is a popular method for collecting questionnaire data.

OMR avoids human keyboard entry, which is prone to mistakes. OMR allows for fast data entry of a high volume of data. Forms have to be carefully designed and filled in as marks that are entered in the wrong place will either be ignored or cause the form to be rejected. Care needs to be taken to keep the forms uncreased, as bent or damaged cards are likely to be rejected by the reader.

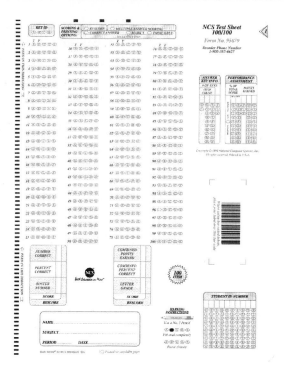

Figure 11.1 Class registation using OMR

Figure 11.1 shows how class registration in a school can be captured using an OMR form.

Optical Character Recognition (OCR)

Optical Character Recognition is a method of data capture where the device recognises characters by light sensing methods. These characters are usually typed or computer printed, but increasingly devices are able to recognise handwritten characters. Many OCR systems used today involve a two-stage process. First a page of text is input using a standard flat bed scanner and the data is encoded as a graphical image. Special OCR software is then used to scan the image, using special pattern recognition software where groups of dots are matched against stored templates. A character is selected when a sufficiently close match is found.

Documents stored in character form can then be input into a word processor. OCR software is becoming increasingly successful with a high rate of accuracy in distinguishing between different characters. Those characters that cannot be matched are highlighted by the software so that the user can make appropriate corrections.

There are problems that arise when OCR software is used to convert scanned images into text. The typeface or handwriting has to be clear otherwise conversion to the correct character may not take place. Problems will also arise if the document has been folded or is creased.

For all its shortcomings, OCR provides a fast way of entering text to be stored in editable form without the need to retype.

Traditionally, OCR has been used for many years to input large volumes of data by commercial organisations such as banks and utility companies. A bank paying-in slip is pre-printed with a bank sort code and customer account number that can be read using OCR.

A **turnaround document** is a document that is produced as printed output from a computer system and then used as input into another computer system at a later time, perhaps with extra data added. An example is the tear-off slip to be found at the bottom of a customer invoice. An individual's invoice will have been printed by a high-speed laser printer together with invoices for thousands of other customers. The top part of the page contains details of the invoice while a perforated section at the bottom is designed to be returned with the payment. Customer account number and other details are pre-printed so that an optical character recognition reader can read the form.

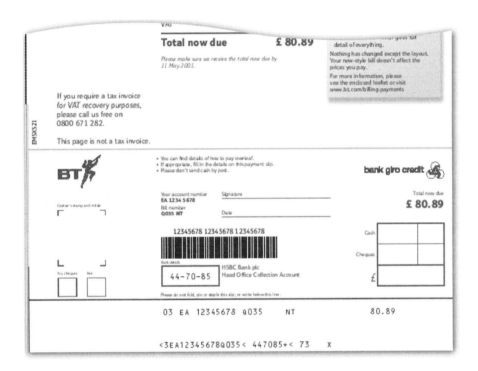

Figure 11.2 Invoice showing turnaround section

Magnetic Ink Character Recognition (MICR)

Magnetic Ink Character Recognition is another fast and reliable method of entering data. Documents are pre-printed with character data in special ink that can be magnetised. The print appears very black. The shape of the characters is recognised by detecting the magnetisation of the ink. MICR is almost exclusively used to read cheques. The bank sort code and customer account number are printed in magnetic ink. The banks' clearing house computers can read these numbers. It is used instead of OCR to help minimise fraud. The hardware is expensive, so MICR is only used in situations when security is important.

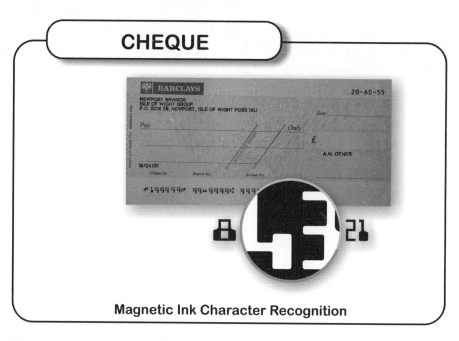

Figure 11.3 A cheque showing MICR characters

Barcode readers (wands and scanners)

Barcodes are used as a means of identification and the pattern of the black and white stripes usually represents a code. Barcodes are attached to objects and have a wide variety of uses. The codes are read from the bars by detecting patterns of reflected light and encoded into computer-readable form using some form of scanner. These devices can be handheld (wand), where the reader is moved over the barcode, or fixed (scanner) where the object is moved across the reader. The barcode has separate right and left sides and can be read in either direction.

Laser scanners are frequently used in supermarkets to read the barcode which identifies the product. Each number in the barcode is represented by four stripes (two black and two white). The barcode is read very quickly with very few mistakes. An article number is a unique number given to a particular product. It is often printed beneath the barcoded representation of the code and is used if, for some reason, the barcode reader fails to read the barcode the operator can key in the code for the product. Two common numbering systems are the European Article Number (EAN) and the Universal Product Code (UPC). These numbers are structured so that each part of the code has a particular meaning. For example, an EAN product code consists of 13 digits. The first 2 stand for the country of origin, the next 5 a manufacturer's number and a further 6 identify a specific product. (The last of these 6 is a check digit – see Chapter 12.)

BAR CODES

Instantly-generated bar codes are used by lottery operators Camelot to identify individual tickets. When the customer's coupon is inserted into the ticket machine, the details are transmitted to the central computer which then allocates an individual transaction code, which in turn is relayed back to the ticket machine which prints a unique bar code. Winning tickets are scanned and verified, and the machine can display how much cash the customer has won.

A bar code generally does nothing more than identify the product. The up-to-date price information is usually stored in a central computer. When an item is scanned the product code is checked against the database which generates a price at the till. One month a special issue of The Banker magazine was priced at £9.95, rather than the normal price of £7.50. Unfortunately WH Smith's database was not updated so when a magazine was scanned at the checkout, the price came up at the lower price. This was a costly mistake for W.H.Smith.

Hospitals use bar codes for tagging new born babies so that no unfortunate mix-ups can occur. Bar codes are also used to identify both books and borrowers in a library. Delegates at trade fairs are often given an ID badge with a bar code,so that exhibitors can scan delegates' badges, thus collecting vital contact details

It is 20 years since bar codes were first used in British Supermarkets. They are now being used in many different ways.

When shopping from home finding the product you want to order from a very long list can be frustrating. Instead, it is planned that consumers with portable scanners can scan the bar code on the tins and packets and so on, when they are nearly empty. The information would then be transferred to a PC and e-mailed off to the supermarket. Small, pen-sized scanners to be used with home shopping catalogues that display bar codes alongside pictures of goods are also being developed.

Couriers use bar coded stickers to track the progress of documents from despatch to signed-for receipt. Portable scanners are used on a package as it is delivered. The data is then transmitted to the central system. This also allows customers to track their package's progress, via the Internet.

Organisations that have to deal with very large amounts of paper, such as the Inland Revenue and the electricity boards, use bar coded tax forms and bills to allow office staff to call up the correct customer details when necessary. Such a system can also be used to improve security: access to an account can be denied unless the appropriate bar code is read.

Considerable use of bar codes is made in the manufacturing industry. Here bar codes are used to identify components, to track goods in the warehouse, and to co-ordinate the delivery of the right parts at the right time along an assembly line.

Figure 11.4 Uses of a barcode

A magnetic stripe card has a strip of magnetic material on the back of a card that is usually made of plastic. Encoded information stored on the magnetic strip on the back of cash, credit and debit cards can be automatically read into a computer when the card is 'swiped' across a reader. The pattern of magnetisation is detected and converted into encoded binary form for computer use. This process is much quicker than typing in the card number and is accurate, not being susceptible to human error. The stripe can be damaged if the card is exposed to a strong magnet as the magnetic pattern would be altered and so render the card unreadable.

Magnetic stripe cards are now widely used on train tickets. The stripe will hold data concerning the proposed journey. On the London Underground tickets are read when a passenger arrives at or leaves a station. As well as being used to confirm the validity of the ticket, data is collected on all the journeys made. This can be used to produce information on passenger patterns and flows that can be used when planning services.

Figure 11.5 London Underground ticket

When parking a car at Heathrow airport, drivers are issued with a ticket that has a magnetic strip on the reverse side. At the time of issue, the date and time of arrival are encoded on to the strip. When leaving the airport building to return to the car park, the driver inserts the card into a large payment device. The device has a magnetic stripe reader that reads the data from the card. The charge, based on the length of stay, is calculated and displayed on a small screen. After the driver has made the payment, using cash, credit or debit card, details of the payment are recorded on to the stripe on the card which is then returned to the driver. When the driver leaves the car park in his car he inserts the updated ticket into a device controlling the barrier.

Data loggers

Another example of data capture in use is to be found in the electricity companies who provide meter readers with **hand-held data loggers**. These are small computers little bigger than a pocket calculator, with a key-pad for input. Meter readings are entered into the data logger as the meter is read. Later on, the data logger is connected to a computer and the data down-loaded.

Sensors

Data can be collected in such a way as to be automatically entered into the computer. For example, computers can be connected to sensors that monitor changes in physical attributes such as temperature, humidity, light, pressure or wind speed. Data from the sensor is often in analogue (wave) form. This has to be converted into a digital (binary) coded form for computer use.

Examples of uses of computer sensors for data capture include:

- a traffic count where pressure pads are used, as changes in pressure are created when a car drives over a pad

- an automatic weather station, particularly in remote and inhospitable places, where different sensors can be used to capture samples of properties such as temperature, humidity and wind speed and encode them in digital form

- timing devices in swimming or athletics races

- temperature sensors to allow computers to control central heating

- PIR sensors for burglar alarms.

Touch Tone telephones

A method which is being increasingly used for limited data capture is a touch tone telephone. Banking systems allow the user to enter her account number using the keys on the telephone handset. In response to voice commands, the user can choose the type of transaction required by press a specific key.

Touch Tone entry is commonly used for the routing of calls for large institutions and simple information systems. For example, when telephoning a cinema to book seats for a particular film the following dialogue could occur:

Computer

> *Good evening, this is BigScreen Cinema. Please press the key to make a choice.*
>
> *Press 1 if you wish to find out about our current program.*
>
> *Press 2 if you wish to book seats.*
>
> *Press 3 if you want information on forthcoming attractions.*

User enters 2

Computer

> *Choose the number of the film you wish to see.*
>
> *Press 1 for Toy Story*
>
> *Press 2 for Planet of the Apes*
>
> *Press 3 for Titanic*
>
> *Press 4 for Bambi*
>
> *Press 5 for Traffic*

User enters 3

Computer

> *Choose the day you wish to attend:*
>
> *Today – type 19*
>
> *Tuesday 20th – type 20*
>
> *Wednesday 21st – type 21*
>
> *Thursday 22nd – type 22*

User enters 21

And so on...

Speech recognition

Speech recognition is a growing area of computer input. As computers get faster and memory increases, reliable speech recognition has become cheaper. Already mobile phones can recognise the name of the person the user wants to call. Soon we will be able to use speech technology to access information without the need for a keyboard or a mouse.

There are two main uses made of speech recognition systems. They are used to enable large quantities of text data to be input as words into a computer so that they can be used in a software package such as word processor. Alternatively, speech recognition systems can be used to input simple control instructions to manipulate data or control software. This is particularly appropriate when the environment makes the use of keyboard or mouse unsuitable, such as in a factory. Users of such systems can create and modify documents in a word-processor; enter data into a variety of ways including form filling. The user can surf the Web or send e-mails simply by speaking.

Speech recognition systems provide users who have a limited ability to type, possibly due to repetitive strain injury or other disabilities, with a manageable way of interacting with a computer system.

Speech recognition systems are used to enter large volumes of text and are particularly useful for professionals who are not skilled in keyboard data entry. In the past such users would have dictated to a secretary or used a dictaphone so that a typist could later audio type the text. The use of speech recognition software allows the user to input the text directly without the need for a typist. This removes any time delay and provides the user with control.

Advantages of using a speech recognition system

Smaller, cheaper systems cannot easily interpret normal speech where words tend to run into one another. The user must speak each word separately leaving clear gaps between the words. Systems need to be trained to recognise an individual's voice.

- Users who are not trained typists can achieve faster data entry than through keyboard use.
- Leaves hands free for other purposes.
- Avoids health risks such as RSI.
- Enables disabled users to enter data.

Limitations of speech recognition practice

- The software needs to be trained to recognise an individual voice.
- Excessive background noise can interfere with interpretation of speech.
- Words that sound the same can be confused (such as their, they're or there).
- Many cheaper systems only recognise a small number of words.

Case Study 1

Dragon NaturallySpeaking® Medical 5.0

(An extract from an advertisment for a specialist voice recognition system for the medical profession.)

Dragon NaturallySpeaking® Medical 5.0 is a fully customisable, powerful productivity solution that enables health care professionals to streamline their workflow using speech. Users can create patient notes and letters, enter data, assemble and manage documents, create and fill-out forms, send e-mails, browse the Web, retrieve information, launch applications and perform other complex tasks by speaking. Users can dictate patient notes and documents directly into a PC or hand-held recorder.

Key Features:

○ Increase Productivity – Dictate memos, letters, reports, email and more as fast as you speak!

○ Get started quickly – Teach the computer to recognize your voice in as little as 5 minutes.

○ Large Medical Vocabulary – Includes a comprehensive 250,000-word vocabulary, including specialised medical terms and phrases. It's easy to add new terms.

○ Create custom voice commands (macros) – A powerful feature that enables you to create documents, insert boilerplate text, launch applications, fill-in forms, and perform other complex tasks with a spoken word or phrase.

○ Work on the Go – If away from your PC, use with an approved hand-held recorder.

○ Listen to Your Work – Hear text, such as an incoming email, read aloud with text-to-speech technology.

○ Surf the Web by Voice.

○ Describe how a speech recognition system could be a useful tool for a dentist.

Key-to-disk entry

Where large volumes of data have to be entered, off-line key-to-disk systems may be used. In these systems the data entry clerks type in the data, which is stored straight away on disk. This method would be used in batch processing (see Chapter 16), where large batches of data are entered before being processed. (As the entry is off-line, it does not affect the performance of the main computer.) Key-to-disk data entry involves the transcription of data that has earlier been captured on paper.

Direct Data Entry (DDE) using a keyboard

Direct data entry is the input of data directly into a computer system for processing immediately. Here the data could have originally be captured on paper, or could be based on verbal data from a customer or over the telephone. Direct data entry is a very widely used method of data entry. However, the speed and accuracy of the data that is input depends upon the skill of the operator.

Activity 1

Explore the methods of data capture in several shops in your local area and produce a table in the format given below. Try to look at as many varied types of shops as possible. Consider all kinds of data that is captured (products sold, stock, payments).

Shop	Data being captured	Method	Reasons for choice

Case Study 2

Data Capture in Schools

Many schools no longer use the old manual method of registering pupils by putting a mark in a book.

St Joseph's High School uses an OMR system. Pupil names are automatically printed on an OMR sheet. The teacher taking the register puts a mark in the right place. Registers from all the classes are then taken to the school office where the OMR sheets are batched together and read by an OMR reader, which automatically inputs the data into the computer. The computer records lateness and absences, and can calculate statistics.

The mode of operation for this system is a form of batch processing (see Chapter 16).

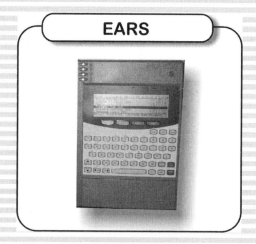

EARS

Figure 11.6 EARS register

Kentland Sixth Form College uses a magnetic swipe card system. Every student has a swipe card and there is a swipe card reader in every classroom. At the start of each lesson, each student swipes their card through a magnetic swipe card reader – these are located near the classrooms and are linked by cable to the central computer.

Woodland Brook High School uses the EARS (Electronic Attendance Registration System) system. Each teacher carries a small hand-held device. In each block of classrooms is a radio antenna which sends and receives messages to and from the teachers' computers. The teacher completes an electronic register on the computer for every lesson. It also can be used for other administrative tasks, such as sending a message to a teacher and storing pupil marks.

The systems used by Kentland Sixth Form College and Woodland Brook High School are both on-line systems.

○ Describe the advantages and disadvantages of each of the three systems described above.

Activity 2

Copy and complete the table given below.

Method	Brief Description	Typical applications	Advantages of use	Disadvantages of use
Key to disk				
DDE				
OCR				
OMR				
Voice entry				
Bar code scanner				
Sensor				
Magnetic stripe card				
MICR				

Activity 3

ParcelFast is an express mail service that handles several millions of parcels in a year. ParcelFast will collect parcels from source, or they can be taken to a ParcelFast office. There are several hundred offices. There are 12 depots where parcels are sorted as they are routed to their destination. Parcels can pass through several depots on their journey from source to destination. As it is most important that the location of parcels is known at all times the time of arrival of a parcel at a depot must be recorded.

- List the data items that must be recorded for each parcel.
- Describe three different methods of recording the data from the parcels.
- Give the advantages and potential problems of each method.

Summary

There are a number of factors that must be considered when choosing a method of data capture for a particular system. These include the volume of the data, the speed that it has to be input, the nature of system and the necessary ease of use, the state of current technological development as well as cost.

At data capture data is encoded into machine-readable form.

Data capture methods include:

OCR	Touch Tone telephone
OMR	Speech Recognition
MICR	Sound
Barcode scanning	Key-to-disk
Magnetic stripe	DDE
Sensors	

Data capture questions

1. Describe three methods of data capture and give applications for which each would be appropriate. *(6)*

NEAB Specimen Paper

2. Use of voice recognition is increasing steadily. Give three reasons why this is so. *(3)*

3. Newtown is setting up a road marathon race and is expecting an entry of several thousand competitors. They need to capture data that will demonstrate the progress of each runner at several stages of the race.

Suggest three methods of data capture and for each describe its advantages and problems. *(6)*

4. An insurance company wishes to enter the data from hundreds of proposal forms filled in by customers each day.

a) Describe a suitable method of data input.

b) Justify your choice

c) Explain a method used to reduce the number of errors made at this stage.

NEAB Computing Specimen Paper 2

5. Speech recognition systems for Personal Computers are now becoming more affordable and useable.

a) State two advantages to a PC user of a speech recognition system. *(2)*

b) Give two different tasks for which a PC user could take advantage of speech recognition. *(2)*

6. Speech recognition systems sometimes fail to be 100% effective in practice. Give three reasons why this is so. *(3)*

NEAB 1999 Paper 2

7. A village store has just installed a computerised point of sale system including a bar code reader.

a) Describe two advantages that the store gains by using a bar code reader attached to a computerised point of sale system. *(4)*

b) Describe one disadvantage to the store of using a bar code system. *(2)*

AQA ICT Module 2 Jan 2001

8. The local council wishes to store the contents of documents on a computer system. The documents consist of hand-written and typed text. The documents will be scanned and OCR (Optical Character Recognition) software will be used to interpret the text and export it to files that can be read by word processing software.

a) Describe two problems that could occur when scanning and interpreting the text. *(4)*

b) Describe two advantages to be gained by using OCR software. *(4)*

c) State three types of material, other than text, that could also be input using the scanner. *(3)*

AQA ICT Module 2 June 2001

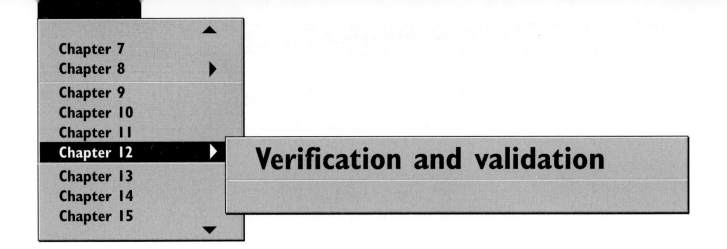

Verification and validation

Errors

Why are errors a problem?

Some errors in data can lead to catastrophic results, whilst others can affect the acceptability of a computer system. As businesses and other organisations are dependent on information from computers, it is essential to try to reduce the occurrence of data errors and so ensure that the information is correct.

Results of errors

There are many examples in newspapers of the results of errors occurring in computer systems. These include pensioners receiving gas bills for millions of pounds, others who are sent letters threatening legal action unless an outstanding payment of £000.00 is received within ten days. A bug caused a newly developed fighter plane to turn upside down when it crossed the equator. (Luckily this was detected during simulation testing.)

When errors can occur

Errors can occur at several stages in a system:

- when the data is captured
- when data, initially collected on paper, is transcribed (copied), usually via a keyboard, into a computer
- when data is transmitted within a computer system, particularly over a network link
- when data is being processed by software.

Activity 1

Research, using the Internet, recent examples of computer errors. Try to find six examples. You will find newspaper sites a good source. Summarise your findings in your own words.

Errors in data capture

The majority of errors are introduced at data entry. Consequently direct data entry methods are used whenever possible. Most of the methods of data capture described in Chapter 11 are reliable and automatic. For example, by using bar code readers at a supermarket check-out, no data needs to be typed in, thus saving time and removing the chance of transcription errors. However no method of data entry can guarantee to be error-free.

It is not always possible to use automated data entry methods. As stated above, at a supermarket check-out, the use of barcoded product codes which can be read by a scanner removes the need to key in data. However, in many supermarkets fresh products such as vegetables are weighed at the check-out and the appropriate code is keyed in. If the operator wrongly identifies the vegetable or mistypes the code, then incorrect data is input. In some systems the operator often is given the facility to enter a number for multiple items rather than scan each item separately, as miskeying could result in a customer being charged for 50 tins of baked beans instead of 5!

Errors in automatic data capture can occur if the system has not been set up correctly. For example, the wrong race times would be recorded at an athletics event if the timing device did not start immediately at the start of a race.

The data for many systems originates on paper. Application forms of many types, mail order requests, census forms and details of car repairs undertaken are all examples where data is often first captured on paper. Errors can be recorded at this stage. A common mistake is to enter a date of birth with the current year. Forms that do not clearly indicate what is required will lead to errors.

Transcription errors

Whenever data is manually copied there is a chance that errors will be made. These copying errors are known as transcription errors. They can occur when the person involved misreads what is written or mishears what is said. Poor handwriting and unclear speech over the telephone are both likely to lead to transcription errors. Long codes made up of numbers or letters that have no particular meaning to the person keying in the data are particularly vulnerable to error. The skill and accuracy of the typing of the person inputting the data will have a major effect on the number of transcription errors made.

Activity 2

For each of the transcription errors below explain what is wrong and why the error is likely to have been made:

- SO23 5RT entered as S023 SRT
- Leeming entered as Lemming
- 419863 entered as 419683
- 2000000 entered 200000
- Hatherley Road entered as Haverly Road
- 238.591 entered as 2385.91
- 23/5/89 entered as 23/5/01
- 199503 entered as 195503

A common form of transcription error is a transpositional error where the order of two characters is mistakenly swapped. A code number 134638 could easily be mistyped as 136438. It is likely that this code number is being used to identify something such as a bank account, a hospital patient or a product. Entering the wrong code could have serious consequences, resulting in the wrong person's bank account being debited, the wrong patient's notes being updated with a drug prescription or the wrong product being ordered for a customer.

Transmission errors

Data that has been entered correctly in a system can become corrupted when it is transmitted within a computer or when sent from one computer to another. This may be due to a poor connection or 'noise' on the line. All data is held within a computer and transmitted in binary coded form – everything is represented by a string of ones and zeros. Characters are most commonly stored in ASCII format and there are coding systems for storing other types of data such as graphics, numbers and sound. Data corruption will result in one or more zeros being written as ones, or ones being written as zeros. It is likely that the resulting string of bits would be interpreted as another character, graphic etc.

For example, if just 3 bits were corrupted when transmitting the characters ICT in ASCII the result could become HGV

1001001 (I) 1000011 (C) 1010100 (T)

1001000 (H) 1000111 (G) 1010110 (V)

Processing errors

Errors occur due to incorrectly written software. Experience shows that programmers produce 30 – 100 faults in every 1000 lines of code. Calculations could be worked out incorrectly, the wrong record could be updated in a file or certain types of transactions could be ignored.

Errors can also occur due to omissions in the specification (the document that lays out what a new system is to do), where certain situations are not considered when the system is designed. Design must always assume that an operator can make errors. One computer system tried to raise the temperature in a chemical process to 800 degrees when the operator had meant to enter 80, with the result that a poisonous chemical was released into the atmosphere. A fatal error occurred when a patient received a lethal dose of radiation treatment for a cancerous tumour. The computer-controlled equipment had previously been used hundreds of times without causing an error. On this particular occasion the operator had altered the data entered by using the cursor keys in a particular way that had not been anticipated.

Activity 3

Discuss possible *causes* of the following: (there may be a number of possibilities):

- a tenant is sent a letter that states that rent is overdue when in fact it has been paid in full

- an electricity bill demanding a payment of 2p is sent out

- a customer receives a copy of 'Bridget Jones' Diary' instead of 'Gladiator' from HOME VIDEO, a mail order company.

Discuss the possible *consequences* of the following errors:

- the wrong patient identification number is entered when entering the results of a blood test

- an electricity meter reading is wrongly entered

- 37 is entered instead of 73 as a student's mark in an ICT module exam.

Reducing errors

Obviously, every effort has to be made to minimise the number of errors that can occur in a computer system.

Reducing data capture and transcription error

The use of automatic data entry, where there is no need for the keying in of data, removes the possibility of transcription errors. In a computerised system for a lending library, a hand-held scanner is used to read the reader's identification number from a barcode on their membership card as well as the accession number which is held on a barcode in the book.

Methods of recording the current meter reading of a householder's electricity have developed through the following systems:

○ A form filled in by the reader that was later transcribed using key–to–disk

○ A pre-printed OMR card where appropriate boxes were shaded to represent the digits

○ A hand held computer with data relating to customer's account so that data can be entered and checked at the house.

Activity 4

For each of the following systems, suggest how a method of data capture could remove the need for data transcription.

○ In a school library, the name and class of pupils together with the book accession number and return date are keyed in whenever a pupil borrows a book

○ Lists of the marks that pupils were awarded in their AS level course work are sent in to the examination board from centres. On receipt of these lists the marks are keyed in by administrative assistants.

○ Employees in an office write down the time of their arrival and departure in a book. These figures are keyed in using DDE at the end of the week so that overtime pay can be calculated.

○ Members of a postal book club can order books by telephone. A clerk takes details of members' account numbers and the codes of the books required then keys in the data using a DDE system.

The use of a turnaround document (see Chapter 11) that can be read in using an OCR reader minimises the amount of data that has to be keyed in.

As a transpositional error is a very common form of typing error, it is crucial it is avoided as far as possible. Paper forms need to be designed with great care so that the chances of errors being made are kept to a minimum.

A data entry screen should be designed to follow the layout of data on the paper form. Mistakes are more likely to be made if the eye has to jump around the page to find an appropriate data field.

The greater number of characters that need to be keyed in, the greater the chance of error. Wherever possible codes should be used.

To ensure that only valid choices can be entered, options can be given for the user to choose, either in the form of tick boxes or drop-down lists.

Human errors such as transcription errors can be detected using two techniques – **validation** and **verification**.

Verification

Verification is used to check that data is entered correctly. The most common method of verification involves typing data into the computer twice. The computer automatically compares the two versions and tells the user if they're not the same. When network users change their password, they have to type in the new password twice to verify it. The key-to-disk method, where data is entered and stored on to a magnetic disk off-line before it is entered into the system as a batch, can allow the data to be entered twice for verification.

Validation

Validation is computerised checking that detects any data that is not reasonable or is incomplete. There are many different validation techniques and one or more might be appropriate for a particular data item. The most common techniques are listed below.

- **Presence check**. This checks that an entry has been made for the field. For example, the Surname field in an order form cannot be left blank.

- **Range check**. This checks a value to be within in a certain range of values. For example, the month of a year must be between 1 and 12. Numbers less than 1 or greater than 12 would be rejected.

- **Field check (format** or **picture check)**. This checks that data is of the right format, that it is made up of the correct combination of alphabetic and numeric characters. A National Insurance number must be of the form XX 99 99 99 X. The first two and the last characters must be letters. The other six characters are numbers. The total length is nine characters. Any other format is rejected.

- **Cross-field check**. This checks that data in two fields corresponds. For example, if someone's gender is stored as Female, their title cannot be Mr. If the month in a date is 04, the day cannot be 31 as there are only 30 days in April.

- **Look-up list**. This checks that the data is one of the entries in the list of acceptable values. For example, the day of the week must be from the list Monday, Tuesday and so on.

Figure 12.1 Look-up

○ **Check digit**. This is used to check the validity of code numbers, for example product codes in a supermarket or bank account numbers. These numbers are long and prone to data entry errors. It is crucial that such numbers are entered correctly so that the right record in the file or database is identified. A check digit is an extra digit added to the end of the original code. The value of the check digit is determined by the value and positioning of the other digits: for any given code there is only one possible check digit. When code has been entered, the check digit is calculated and compared to the entered value. If the two digits do not match, an error is reported.

Extension Activity

The most common method of calculating a check digit is based on Modulo-11 as follows:

A short identity code number may consist of five digits plus one check digit at the end.

For example, with an ID code of 69247

Multiplying each digit in the code by another number called the weight. Then add them, for example

multiply

first number by 6	(36)
second number by 5	(45)
third number by 4	(8)
fourth number by 3	(12)
fifth number by 2	(14)

Add them up 36+45+8+12+14 = 115

Divide by 11 10 remainder 5

Take the remainder away from 11 gives 6. The check digit is 6.

(**Note**: if the remainder is 0, the check digit is 0. If the remainder is 1, the check digit is X.)

ID code number = 692476

Now, if a transpositional error is made and the code number is mis-entered as 629476 then

multiply

first number by 6	(36)
second number by 5	(10)
third number by 4	(36)
fourth number by 3	(12)
fifth number by 2	(14)

Add them up 36+10+36+12+14 = 108

Divide by 11 9 remainder 9

Extension Activity *continued*

Take the remainder away from 11 gives 2. The check digit should be 2 – but it has been entered as 6: a data entry error has been detected.

○ Work out the check digit using modulo-11 for the following numbers:

453891 776432 678310

Remember: Check digits are only used with codes that are used for identification purposes such as a bank account number, a credit card number or a student identification number.

Case Study 1

Accuracy versus validity

Validation checks can ensure that the data entered is reasonable and sensible, that it obeys set rules. It does not mean that the data is correct. It is possible that data entered could be valid but inaccurate. A temperature sensor recording the temperature in a furnace may give valid data, a number within a certain range, but the data may not accurately reflect the temperature if the sensor is not set up correctly. If 25 The Glebe, the address of a customer, is entered as 52 The Glebe or even 61 Brockhampton Road, no validation error would be detected as both are reasonable addresses. However, neither is an accurate address for the customer. A man's date of birth entered as 25/3/1975 is a valid date but would be inaccurate if the man had been born in April.

Which of the following errors could be detected using validation checks?

○ A car registration of 234 B 65

○ A date of birth for a 6th form college student of 12/4/1999

○ An entry of a quantity of 20 tins of dog food instead of 2

> **Validation of meter readings** >

An electricity company reads the meters in individual houses. These readings will be validated by a range check to make sure that they are within a sensible range. The records are fed into the computer.

The computer can check that the details for every house have been entered by checking that a control total – the number of houses – is correct. Of course, one house may have been omitted and another entered twice. This can be checked by making sure that no customer number is repeated.

When the bill is calculated it should be validated by another range check to make sure that it is not ridiculously high or low.

Transmission error checking

Transmission errors can be avoided by the recipient computer sending the same message back to the original sender. If the two messages are the same, the data will be correct. The use of **parity** is a method that enables the detection of errors that can occur when data is being transmitted or stored. Electronic 'noise' can cause the state of bits to be altered – resulting in a 0 being changed to a 1 or a 1 to a 0.

A **parity bit** is an extra bit that is added on to a group of bits. The parity bit is solely used to check that the other bits have not been corrupted. There are two types of parity – even and odd. When even parity is used the parity bit is chosen to be 0 or 1 so that the total number of transmitted bits set to 1 (including the parity bit) is even.

Extension Reading

For example, using even parity:

	Parity bit	total number of 1 bits
0 1 1 0 1 1 0	0	4
1 1 0 0 1 1 1	1	6
0 0 0 0 0 0 0	0	0

When **odd parity** is used the parity bit is chosen so that the total number of 1 bits is odd.

For example using odd parity:

	Parity bit	total number of 1 bits
0 1 1 0 1 1 0	1	5
1 1 0 0 1 1 1	0	5
0 0 0 0 0 0 0	1	1

When the data is received the bits are checked and if the parity is incorrect then one or more of the transmitted bits has been corrupted and appropriate action can be taken.

More complex forms of checking

The use of a parity bit will detect if a single bit has been corrupted but it will not identify which bit. If the specific bit could be identified, then it could be corrected. The use of **vertical parity checking** enable this. Using this technique a normal parity bit is transmitted with every word. (This is sometimes called **horizontal parity**.) After a set number of words have been transmitted, an extra word is sent which merely used as a check. The setting any specific bit in this word is determined by using the rules of parity on the bit in the same position of the preceding words. The example shown below shows the use of vertical parity.

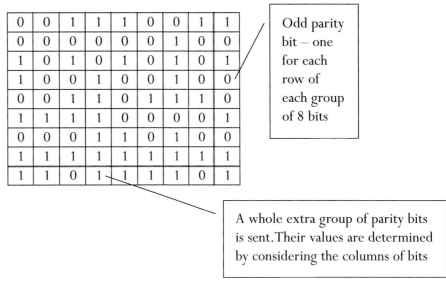

Odd parity bit – one for each row of each group of 8 bits

A whole extra group of parity bits is sent. Their values are determined by considering the columns of bits

Figure 12.2 Parity

When the parity word is read, the setting of each bit is checked by recalculating the parity bits. A bit that has been corrupted can be pinpointed through the use of the two forms of parity.

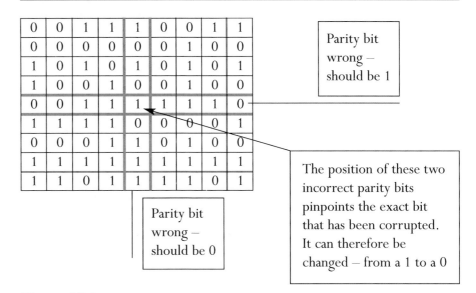

Parity bit wrong – should be 1

Parity bit wrong – should be 0

The position of these two incorrect parity bits pinpoints the exact bit that has been corrupted. It can therefore be changed – from a 1 to a 0

Figure 12.3 Parity

> **Errors in processing: testing**

Processing errors caused by incorrect coding should be spotted at the system's testing stage. Testing should not just include simple test data but should include extreme data (such as large numbers) and incorrect data. Results should be compared with expected results. Incorrect data should be rejected. Testing may need to be performed on different platforms, for example computers with different memory sizes to make sure it works on all machines.

Errors in processing: specification, analysis and design

A very detailed and rigorous specification and analysis is required to ensure that all possible situations are taken into account. If this is not done, the programs that are produced will not take into account all combinations of data entry and will therefore produce errors at under certain circumstances. There is a range of techniques that can be used to help to produce a specification that is clear and unambiguous. It is most important that the design and subsequent implementation follows the specification absolutely.

Particular care should be taken if software is being adapted to a new situation. When the Air Traffic Control system at Heathrow was updated ten years ago, software that had been developed for airports in the US was installed. Very soon some strange effects were noted – planes on the screen seemed to flip over! After investigation, the cause of the error was found to relate to the Greenwich Meridian. No account had been made in the software for the fact that part of the UK lies west of the meridian and part east. All locations in the US have values between 65° and 170° west. Leicester and Norwich lie at a similar latitude; Leicester has a longitude near to 1° west while Norwich lies near to 1° east. As the software did not distinguish between east and west values the two locations were taken to be the same. Fortunately the fault was detected during testing.

File totals

In batch processing, transaction and master files are read from start to end. An extra record is usually stored at the end of such a file and is used to check that data has not been corrupted.

As records are written to a file, running totals of certain fields are built up. These are called control totals. At the end of the file an extra record that contains the control totals is written. When the file is re-read in a later process the running totals are built up as the records are read. When the final record is read the running totals are compared with the control totals. If they are not identical an error has occurred and the software needs to take appropriate action.

Examples of fields that could be used as control totals:

- total number of records in the file

- total of current stock created by adding up the stock level field for each item

- total value of deposits in a transaction file for as banking system.

Sometimes totals are kept that have no meaning but are included only for error checking purposes. Such totals are called hash totals. An example for a suitable field for a **hash total** would be a product identification code field.

Record locking

In a transaction processing system records are updated as a transaction occurs. In a shared system, where there is more than one user, the situation could arise when two users were updating the same record simultaneously. For example, imagine that two travel agents from different offices were trying to book passengers on to the same flight at the same time. Both would read the record for the flight – which indicates that four seats remain. Travel agent A carries out a transaction to book two seats for her clients and the updated record is written back, with two seats now available. Immediately afterwards travel agent B rewrites the record (which showed four seats available) and carries out a transaction to book three seats for the new clients. When the updated record is written back it overwrites the record that travel agent A updated and her transaction is lost.

To ensure that such errors can not occur the software will be written in a way that prevents two users accessing the same record simultaneously. This is usually done by **locking** a record until a transaction is complete.

Back-up (see Chapter 19)

It is not sufficient to find out that an error has occurred – it is necessary to recover from that error. An effective and thorough back-up regime is vital to ensure that, if a file is corrupted or lost, it can be created in its original, correct form.

File version checks

Errors could occur if the wrong version of a master or transaction file were to be used by mistake. Consider a system in use by a credit card company. The system maintains a master file that holds details of the state of each customer's account. A transaction file made up of records each of which hold the details of a debit to an account is used to update the current amount owed field in the master file record. If, by mistake, yesterday's transactions were used instead of today's, the same debits would be made twice from an account. Data files will contain a header record that will contain data items such as version number and date created. These would be checked at run time to ensure that the correct version of a file were being used.

Report generation

Further checking should take place when reports are to be generated to ensure that the information is sensible. Regular utility bills such as gas, electricity or telephone can be checked against previous bills for the same customer and any major discrepancies highlighted. The total owing on bills can be checked to be within acceptable limits to ensure that bills for very large or very small or even negative amounts are not sent out to customers.

It is important that the date and in some cases the time is included on all reports so that the information is not wrongly interpreted.

Activity 5

Newtown 6th Form College
Application form 2002

N

Please complete the form in black ink using BLOCK CAPITALS

Surname	
Forename(s)	
Home Address	

Telephone number	Post Code
Date of Birth	Age on 1/9/2002 yrs mths
Previous School	

Ethnic origin (please tick)

White ☐   Black Caribbean ☐   Black African ☐   Black Other ☐   Indian ☐

Pakistani ☐   Bangladeshi ☐   Chinese ☐   Other ☐   Info refused ☐

Provisional Course for September 2002
Please indicate which courses you are interested in. Please list, in order of preference, the subject name and the level (AS, A2, AVCE, GCSE, GNVQ Intermediate or Foundation).

Subject	Level

Current Courses of Study
Please list the GCSE subjects that you are currently studying, or have already taken, indicating which level paper you are being entered for (higher intermediate or foundation). Please ask your school to add your predicted (or actual) grades.

Subject	Level	Predicted Grade

Figure 12.4 Application form

Study the application form for Newtown 6th Form College shown in Figure 12.4.

Determine the validation checks that would be necessary for each field so that as many errors as possible can be trapped.

Summary

The information produced by a system is only as good as the quality of the data that is input.

Errors in data can have far-reaching effects.

Errors can occur when data is

- captured
- transcribed
- transmitted
- processed

Errors can be reduced in a variety of ways including:

- using automatic data capture methods such as OCR or OMR that remove the need for transcription
- verification checks
- validation checks

Verification and validation questions

1. The Driver and Vehicle Licensing Centre want to validate car registration numbers. These are now of the form:

   FN52XYZ

   a) Suggest five different checks which accept all the above which together would reject the following invalid codes, for example : F52XYZ fn52xyz FN523XYZ FN52XY6 4087XYZ

   b) Why might a valid registration number still be incorrect?

2. The date of birth of applicants for jobs is stored on file as a six-figure digit e.g. 120167. Suggest three ways you could validate this data.

3. An insurance company wishes to enter the data from hundreds of proposal forms filled in by customers each day.

   a) Describe a suitable method of data input.

   b) Justify your choice

   c) Explain a method used to reduce the number of errors made at this stage.

   NEAB Computing Specimen Paper 2

4. A school uses an information system to store details of students' examination entries and results. As part of this system a program is used to check the validity of data such as candidate number, candidate names and the subjects entered.

   a) Suggest three possible validation tests which might be carried out on this data. *(3)*

   b) Explain, with an example, why data which is found to be valid by this program may still need to be verified. *(2)*

   NEAB 1996 Paper 2

5. A well designed information system should be able to check that input data is valid, but it can never ensure that information is accurate.

   a) Explain the distinction between accuracy of information and validity of data. Illustrate this distinction with a suitable example. *(4)*

   b) Describe two ways in which data capture errors may arise, together with techniques for preventing or reducing these errors. *(4)*

6. SupaGoods is a home sales company. Catalogues are left at people's homes. A local agent calls two days later to take orders and collect the catalogues. The agent sends the details of the goods ordered to the Head Office where they are processed. The completed order is returned to the agent who distributes the goods and collects payment.

   a) Describe two distinct methods of data capture for the agent. State one advantage and one disadvantage of each method. *(6)*

   b) The orders are validated at Head Office.

   i) Explain what is meant by validation. *(2)*

   ii) Describe briefly two validation checks that might be carried out on an agent's order. *(4)*

   NEAB 1998 Paper 2

7. A company that organizes music festivals sells tickets via the Internet. An on-line booking form has to be completed to reserve the tickets.

   a) State three fields other than Name, Address and Postcode that you would expect to find on the booking form. *(3)*

   b) Name and describe a different validation check for each of the fields that you have chosen *(6)*

   AQA ICT Module 2 June 2001

8. An electricity supply company needs to arrange for householders' meters to be read regularly. Meter readers visit each house and record the current meter reading for each account on a paper data capture document. At the end of the day all the documents are read directly into the computer system to avoid transcription errors.

   a) State two items of data that should be printed on the data capture document before it is given to the meter reader. *(2)*

b) Describe two validation checks that should be performed when the data is read into the computer system. *(4)*

c) How is it still possible for incorrect data to be stored in the computer system? *(2)*

d) Since the company has a large number of accounts, the billing process is run overnight at regular times. Explain why batch processing would be appropriate. *(3)*

NEAB 2000 Paper 2

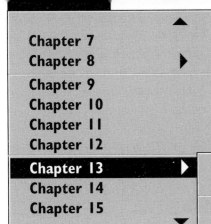

Organisation of data for effective retrieval

All organisations need to store and retrieve data. The data stored and structure is referred to as a **database**. The software package that allows a user to set up and access a database is called a **database management system (DBMS)**. Microsoft Access is an example of a DBMS. A DBMS allows the user to access the data in many different ways.

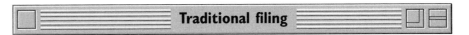

Traditional filing

In the early days of computing, individual applications were developed independently within an organisation. For example, in many organisations the payroll was the first application to be computerised because of the increased speed of producing pay slips compared with manual methods. Personnel records might then be computerised but on a completely different computer system.

The term **flat file** is often used to refer to these single files that are like 2-dimensional tables. A flat file is similar in structure to a manual card file: there is one card (or record) stored for each entry in the file.

At your school or college, when computers were first used to aid administration, a number of applications will have been developed each based on its own flat file. These applications would most probably have included:

- examination entry system
- attendance system
- student record system
- timetabling.

The **examination entry system** would need a file of data relating to students and the subjects they were taking.

The **attendance system**, perhaps using OMR data entry from class registers, would need to access a file of student and class details.

The **student record system** would use a file that held details of a student's previous educational history and basic personal data.

Each separate application had its own data files that were accessed by a number of programs. Although these computer systems brought many benefits over non-computerised methods, there were still a number of problems.

There was a **duplication of data**: fields such as name, address and data of birth were stored in different files for each of the systems. Unnecessary data duplication is known as **data redundancy**. Redundancy often leads to **data inconsistency**. Inconsistency occurs when the same data item, held in two separate files, is different. This could perhaps occur when data is originally entered into the systems (date of birth could be typed correctly into the record system but incorrectly into the examination entry system) or if a data item were changed in one system but not another (a student changes home address – the amendment is made in the student records system but not the examination entry system).

The data files used in each system were linked very closely to the programs. Any changes to the structure of the file for example if an extra field, such as previous school, needed to be added to a student's record in the student record system, then every program using the file had to be modified.

Data in one system was not available to another system and if information was needed in a different format then a new program would have to be written.

Using a DBMS (Database Management System)

More and more complex filing methods were developed and from these developments database systems were evolved. In a database system data is kept separate from the applications programs that use it: the DBMS works between the two. Ideally, each data item is stored only once in the database. Thus inconsistencies of data do not arise.

Your school or college will probably now hold all the required data in one database that will support a wide range of applications. Basic student data need only be captured once when a student joins the college and other data added throughout his/her stay. Timetabling, examination entry and attendance tracking can all be integrated into one system. The DBMS used will allow ad hoc one off, occasional reports to be produced easily as required without the need for complex, time-consuming programming.

Relational databases

A relational database is the most common form of database in use today. There are many database management packages using relational methods on the market, from sophisticated systems such as Oracle and SQL server that are used by large organisations, to systems running on stand-alone computers or over local area networks, such as Microsoft Access, or Filemaker Pro.

> **Elements of a relational database**

To build an application using a relational database you need to design **tables** and the **relationship** or links between them, **forms** that are used for data entry and **reports** to present data. **Queries** are used to select data from the database and **macros** can be written to automate processes.

Tables

A table relates to a single **entity**: an object, person or thing. It is made up of **rows** and **columns**. A row represents one instance of an entity and a column holds one data item or **attribute** about an entity. Figure 13.1 shows a table that holds details of CDs in someone's collection. Each row contains details about one CD and each column relates to a specific item of data such as Title or Artist.

CD Code	Title	Artist
CD027	Joshua Tree	U2
CD033	Mutations	Beck
CD045	You've Come a Long Way, Baby	Fat Boy Slim
CD046	Ray of Light	Madonna
CD055	Urban Hymns	The Verve

Figure 13.1 A CD collection showing tables, rows and columns

Primary key

In most systems records in a table of a database need to be identified individually. A key is a field used to identifying a record. A **primary key** is a unique field or group of fields chosen to identify a record.

For a given table there may be more than one possible field that could perform this function and these are called **candidate keys**.

An example of this can be seen in an employee entity that could include Surname, Forename, Address, Employee_Number and National_Insurance_Number as attributes. Employee_Number and National_Insurance_Number are both unique for an employee and are thus candidate keys. It is likely that Employee_Number would be used as the primary key as it is shorter. The Surname is not a candidate key – there could very well be more than one employee with the same surname – and if it were chosen as the primary key all sorts of mix-ups could occur: for example the wrong employee called Jones could be allocated a pay rise! Another problem would be caused by the fact that many women change their surname on marriage. If the surname had been used as the primary key it would have to change, causing all sorts of problems.

In the CD table a field CD Code has been included so that each record can be uniquely identified. There could be two CDs called 'The Hits Album' made by different artists.

Links between tables

Consider a database developed for a veterinary practice. Figure 13.2 shows some of the data items that might be held in a table, this one for pet details in a veterinary clinic.

Pet Code	Name	Type	Date of Birth	Owner Code	Owner Name	Telephone Number
P0123	Misty	Cat	23/1/92	2234	Mary Preston	889976
P0345	Rover	Dog	12/12/90	1995	Julian Giles	765095
P0887	Foggy	Cat	23/1/92	2234	Mary Preston	889976
P1559	Gladys	Gerbil	16/4/98	1942	Amelia Alderson	565643
P1985	Slinky	Tortoise		1995	Julia Giles	765095
P2233	Speedy	Tortoise		1772	Sally Ann Taylor	876875

Figure 13.2 Single table for Pet details at a veterinary practice

The single table shown is not the best way of storing the data as certain data items are repeated. For example, there are instances when one owner owns two pets. Owner 2234 (Mary Preston) owns both Misty and Foggy. Repeating the name and Phone number of this owner for every pet that she owns, has led to unnecessary data duplication.

Owner 1995 owns Rover the dog and Slinky the tortoise. But is this owner **Julian** or **Julia** Giles? Unnecessary data duplication can lead to **inconsistency** where two or more entries for the same data item differ. In this case it is likely that the discrepancy occurred when the data was entered into the database.

The use of a more sophisticated data structure would allow the data to be stored in a way that avoided such unnecessary duplication. Such a structure would store the data in more than one table with links between the tables.

It can be seen that we need two tables to store the veterinary data without unnecessary duplication. See Figure 13.3. The two tables in Figure 13.3 equate to the details of PET and OWNER. Note how the details of Owners 2234 and 1995 now only appear once in the database. The problem of unnecessary duplication has been resolved – but the link between owner and pet has been lost.

Pet

Pet Code	Name	Type	Date of Birth
P0123	Misty	Cat	23/1/92
P0345	Rover	Dog	12/12/90
P0887	Foggy	Cat	23/1/92
P1559	Gladys	Gerbil	16/4/98
P1985	Slinky	Tortoise	
P2233	Speedy	Tortoise	

Owner

Owner Code	Owner Name	Telephone Number
2234	Mary Preston	889976
1995	Julian Giles	765095
1942	Amelia Alderson	565643
1772	Sally Ann Taylor	876875

Figure 13.3

An extra field needs to be added to one of the tables to provide a link.

We could try adding Pet Code to the OWNER table. We could fill in P1559 for Owner 1942 (Amelia Alderson owns Gladys the gerbil) but which code should we enter for owner 2234? She owns both P0123 and P0887 (Misty and Foggy) and there is only space for one code.

We do better if we add the Owner Code to the PET table. Here, as every pet has only one owner, we do not have any problem. Owner Code in the PET table provides the link that is needed. The data item Owner Code is duplicated in the database, but the duplication is necessary. Figure 13.4 shows the two tables linked.

Pet

Pet Code	Name	Type	Date of Birth	Owner Code
P0123	Misty	Cat	23/1/92	2234
P0345	Rover	Dog	12/12/90	1995
P0887	Foggy	Cat	23/1/92	2234
P1559	Gladys	Gerbil	16/4/98	1942
P1985	Slinky	Tortoise		1995
P2233	Speedy	Tortoise		1772

Same field – provides link

Owner

Owner Code	Owner Name	Telephone Number
2234	Mary Preston	889976
1995	Julian Giles	765095
1942	Amelia Alderson	565643
1772	Sally Ann Taylor	876875

Figure 13.4

In a relational database the link between tables is achieved by duplicating the primary key of one table with an identical field in the other table.

In the veterinary example, an extra field Owner Code was added to the PET table to provide the link with the OWNER table. This field is known as a foreign key in the PET table. The format of the **foreign key** should be identical to that of the primary key in the linked table.

> ○ A foreign key is a field that is not a key in its own table but is a primary key in another table.
>
> ○ In relational databases foreign keys are used to enable the links between tables.
>
> ○ The use of foreign keys brings some, necessary, duplication of data.

In the example given, the attribute, Owner–Code, which was used as the foreign key, was already present in the original, single, table. (see Figure 13.2). Very often this would not be the case and an extra field, a primary key for the new table, would need to be created.

VIDEO

Video Code	Title	Category	Loan Price
0981	Babe	PG	2.50
1197	Shakespeare in Love	15	3.50
1280	The Jungle Book	PG	2.00
1543	Antz	PG	2.50
1694	Silence of the Lambs	18	2.00
1769	The Lion King	PG	2.50

MEMBER

Member Code	Surname	Forename	DOB	Phone Number
A7652	Nayyar	Mukesh	12/3/84	832514
A9856	O'Sullivan	Mary	11/4/56	897543
B1100	Price	Paola	5/11/76	675643
B6993	McKenzie	Frazer	23/8/38	877665

Figure 13.5

In some applications, linking tables using a foreign key will not be sufficient. For example in a video hire application where a member can have borrowed many videos and a video can have been borrowed by many different members, the use of a foreign key would not provide the required solution (see Figure 13.5). If Mary O'Sullivan has borrowed three videos

128	The Jungle Book
981	Babe
1769	The Lion King

which of these three codes should be inserted in the foreign key field for Video Code in the MEMBER record? Adding a Member Code as a foreign key to the VIDEO table cannot solve the problem. If the video had been loaned on three successive days to Mary O'Sullivan, Frazer McKensie and Paola Price, then which Member Code should be entered as the foreign key in the VIDEO table?

Creating an extra table that provides the link between the VIDEO and MEMBER tables solves the problem. This table could be called LOAN and might contain attributes, Member Code, Video Code, Date Of Loan.

The two attributes Members Code and Video Code would together form the primary key of the LOAN table see Figure 13.6.

VIDEO

Video Code	Title	Category	Loan Price
0981	Babe	PG	2.50
1197	Shakespeare in Love	15	3.50
1280	The Jungle Book	PG	2.00
1543	Antz	PG	2.50
1694	Silence of the Lambs	18	2.00
1769	The Lion King	PG	2.50

MEMBER

Member Code	Surname	Forename	DOB	Phone Number
A7652	Nayyar	Mukesh	12/3/84	832514
A9856	O'Sullivan	Mary	11/4/56	897543
B1100	Price	Paola	5/11/76	675643
B6993	McKenzie	Frazer	23/8/38	877665

LOAN

Member Code	Video Code	Date Out	Date In
A9856	1280	22/1/99	23/1/99
B1100	1769	24/1/99	1/2/99
B6993	1694	4/2/99	5/2/99
A9856	0981	12/3/99	14/3/99
A7652	1543	23/3/99	24/3/99
A9856	1769	6/4/99	7/4/99

Figure 13.6

Forms

A form is a way of displaying data from a table. A form is usually used when new data is being entered. A DBMS package such as Microsoft Access allows you to design a form for a particular purpose. Figure 13.7 shows a simple form designed in Access for a PET table for a veterinary practice. Many extra features, such as colour and clip art can be included to enhance a form. Figure 13.8 shows a more complex form.

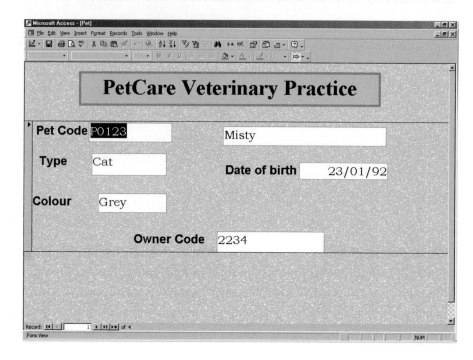

Figure 13.7

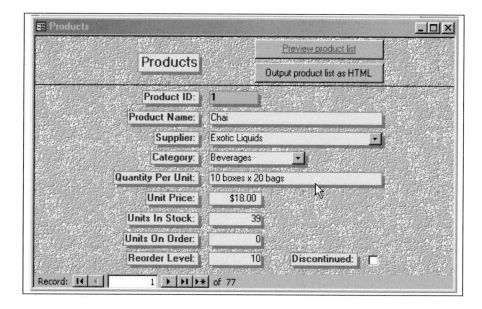

Figure 13.8

Reports

A report is a printed representation of information extracted from the tables of a database. Reports can be set up to include selected data fields, they can show details of many items of data or summarise information using totals; complex calculations can be carried out. The design of the report determines exactly what is included. An example of a report is shown in Figure 13.9. Reports can be designed based on tables or on queries that are used to select and link data.

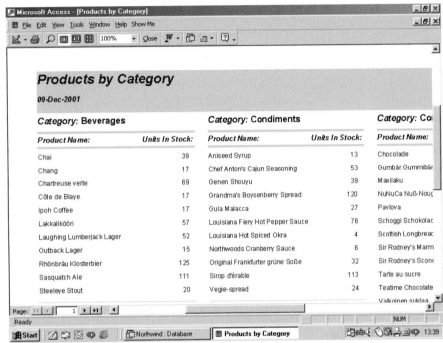

Figure 13.9 Report

> **Queries**

Information retrieval is the extracting of useful information from large amounts of data. The ability to query data in a variety of ways is one of the features that makes a DBMS such a powerful tool. Interrogating the database, searching for data or selecting data are different ways of describing a query. A query is a question asked of the data in a database. A query should be designed so that it produces meaningful information in a way that is appropriate for the user and situation.

Examples of queries:

- Which videos are by Madonna?

- How many cats belonging to Mary Preston are registered at the veterinary practice?

- Which videos are overdue, and who has borrowed them?

●   Produce an alphabetic list of all members who have not borrowed a video in the last month.

●   Which category of video has been the most popular this week?

Queries can be used to view, change or analyse data in different ways. The most common form of query is a select query. This is used to select data from one or more tables by using chosen criteria. The data is then displayed in a form that is specified in the query. Queries can also be used for the source of records for forms and reports.

A query can be used to:

●   search for information in one table

●   search for information in related tables

●   combine the information from related tables into a new table

●   choose which fields are to be shown in the new table

●   specify criteria for searching on; for example find the names of all videos borrowed today.

The query can be saved so that it can be executed when needed.

Programs such Microsoft Access allow queries to be set up in two ways. They can be written in SQL (Structured Query Language), which is a special language for querying the database, or they can be produced using a graphical interface called Query By Example (QBE). Access stores queries produced by QBE in SQL form. There may even be a wizard to help write the query using QBE.

Using QBE involves choosing the fields to be output, establishing the necessary relationships and specifying the criteria to select the appropriate records. Procedures such as sorting or grouping the data can also be specified. Using the query by example method is much easier that programming in SQL, but some queries cannot be done by QBE and SQL has to be used. Queries can be developed by amending the SQL script produced by a QBE.

Figure 13.10 shows a query set by QBE in Microsoft Access. This query searches information from the Pet and Owner tables of the Vet database. It selects all the pets of type "Gerbil" and outputs their names, together with the name and telephone number of their owners. Figure 13.11 shows this.

The SQL script generated by this query is shown below:

```
SELECT Pet.Name, Owner.[Owner Name], Owner.[Telephone Number]

FROM Owner INNER JOIN Pet ON Owner.[Owner Code] = Pet.[Owner Code]

WHERE (((Pet.Type)="Gerbil"))

ORDER BY Owner.[Owner Name];
```

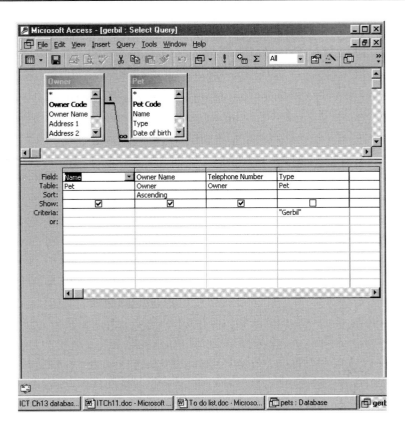

Figure 13.10

	Name	Owner Name	Telephone Number
▶	Gladys	Amelia Alderson	565643
	Horatio	Julian Giles	765095
✳			

Figure 13.11

It would be very tedious and wasteful to have to create a different query for every type of pet. The use of a parameter query allows the specific pet type to be entered at run time. A **parameter query** is a query that when run displays a dialog box that prompts the user for information, such as criteria for searching records. If the 'Gerbil' query shown on figure is changed so that in the Criteria row and Type column the 'Gerbil' is replaced with [Enter Type of pet] (the square brackets are vital here) then the dialog box shown in Figure 13.12 is displayed. When Gerbil is entered (or *cat* or *dog*) then the query is run using 'Gerbil' as the criteria

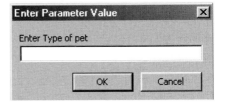

Figure 13.12 Parametes query

Producing information is the prime purpose of a database, so query facilities are fundamental to an efficient DBMS. Structured Query Language is a data manipulation language developed for databases. The main constructs of SQL are select, which specifies which fields are required and join which sets up the links between tables.

Select   (fields required)
From   (name of table)
Where   (criteria for selection eg Certificate = '18')

SQL can be used to write very complex queries.

SQL is a language that is supported by many DBMS.

Sorting data

The information in a database will probably be stored in primary key order. The DBMS can sort the information into any sequence. For example, records may be needed in alphabetical order of surname for one purpose but in chronological order for another. A field that is used to sort data is known as a **secondary key**. This could consist of more than one field, in which case it is called a **composite** key.

Indexing data

Sorting a large database can take a long time. Sorting can be made much faster by creating an index for a field. This keeps a record of the information in the order of this field. If the records of a table are likely to be frequently needed in order of a particular field, then it is worthwhile indexing that field. Indexed fields other than the primary key field are secondary keys.

It is also a good idea to index fields that are often used in a query criteria; for example to find all videos hired yesterday, a query on *Date of hire* would be needed. This field should be indexed.

Note: Indexing a particular field will speed up access to data, but will slow down record up dating since every time an indexed field is changed, the index entry will have to be changed as well.

For a file holding details of video shop members the following fields may be stored:

Surname, Initial, Member_Number, Phone_Number, Date_of_Birth

The Member_Number is likely to be the primary key as it is unique for each member and identifies which member the record is referring to. Date of birth could be a secondary key used for sorting the data into age order.

Macros

A macro is a set of one or more actions that each perform as specific operation such as displaying a form, carrying out a query and printing a report. Macros are used to automate important actions and can be stored for repeated use. Buttons can be set up on the screen so that when a user selects the button the macro is run (see Chapter 15).

Organising a database

Considerable care must be taken when designing and setting up a database. A database design will depend on what needs to be stored and how it is to be accessed. Planning is essential as it is difficult to alter a design at a later date. When a practitioner designs a database system, he must:

- decide on the tables in a proposed database system

- identify the relationship between these tables

- define the structure of each individual table in the database by identifying the fields that belong to each of the tables

- decide upon the type and properties of each field.

The first three factors may well be arrived at by going through the process of normalisation which will be studied in A2. Normalisation is a process that produces a set of tables that have no unnecessary data duplication. If the tables are not correctly structured, errors can become established in the data and queries can be slow and inefficient. Attention will have to be given to the choice of primary keys, some of which may already exist while other tables may require new fields to be established specifically for this purpose to ensure that each record can be uniquely identified. The database should be set up to maintain referential integrity. Referential integrity ensures that records that are related to others cannot be added, altered or deleted if that action were to create a missing link. When rules of referential integrity are in force, you cannot add a record to a table with a foreign key that does not exist as a primary key in the linked table.

For example, in the veterinary application, a pet record could not be entered into the pet table with an owner code of 5693 if there were no owner record with 5693 as a primary key in the owner table. In the video hire application, a video with code 0981 could not be deleted from the video table if there were a loan for the video 0981 in the loan table

> **Choosing fields (attributes) for a table** >

The name of a field should be chosen to be meaningful and should relate as clearly as possible to the attribute it describes. The actual fields must be chosen with care.

When storing information about people, surname, forename and title are usually stored as separate fields. This

- enables the table to be sorted into alphabetical order by surname (it is unusual to sort by forename)

- allows the name to be put together in different ways for different uses

> <Title>   <Forename>   <Surname> on an envelope
> *Mr*   *Patrick*   *McGowan*
> Dear   <Forename> etc.
> Dear   *Patrick*

Addresses are usually stored as a number of different fields that can be output on different lines.

It is usually more appropriate to store date of birth rather than age, as date of birth, unlike age, does not change.

For each field in a table, the designer will need to specify data type, length and choose other properties. The choice of type will allow the DBMS to store the data in the most efficient way and prevent the user from entering the wrong type of data in to a field.

The data types available in most DBMS

Text	A group of characters. Text is used to store letters and numbers when calculations are not to be carried out. For example, text is the most appropriate data type to store a telephone number, customer number or house number. If a telephone number 02089945643 were stored as a number, the leading 0 would be lost. A designer can specify a size for a text field up to a maximum of 255 characters.
Numeric	The number of decimal places stored can be specified.
Date/time	A number of different date and/or time formats can be chosen.
Currency	Is used whenever money fields are to be stored. No rounding off takes place.
Memo	Used for extended text – the length can vary up to 64,000 characters in Access.
Boolean	Can hold only 2 values: true or False
Counter	A field where the DBMS will automatically insert an increasing value each time a new record is created.

When choosing the maximum length for a text field a balance must be made between saving disk space and allowing for possible long entries. It is not always an easy task to choose the appropriate length for a field. For example, how many characters should be allowed in a surname field? Many surnames have fewer than six letters but other double-barrelled ones are substantially longer. Twenty is a size that is often chosen.

An optional **default value** can be selected. A default value is a value that will be inserted into a field at data entry if the user does not enter one for himself. An example for a default value could be today's date: in most situations the user will not need to change this value, thus saving time at data entry.

Validation checks can be selected for a field; for example a month number must be between 1 and 12. An input mask can be applied to a field to ensure that it satisfies a pre-determined format. For example, in a particular application a membership number might always have to be a letter followed by four digits; a post code must consist of letters and digits in the correct combination.

Advantages of database over flat file

- Reduces data redundancy. Data redundancy means that storage space is wasted as data is duplicated.

- As each data item is held only in one place it therefore only has to be entered once, thus saving time and removing the possibility of it being entered differently (which would have led to data inconsistency).

- There is a centralised pool of data: each data item is stored only once but is accessible to all applications.

- Data consistency is maintained: due to the use of linked tables if an item of data is changed it only has to be changed in one place for all applications.

- Data is independent from the applications. A change in the structure of a database, for example the addition of an extra field to the table, can be made without having to alter every application.

- Ad hoc (one off) queries can be made with relative ease by users to meet specific and unforeseen needs. More complex queries can be saved and re-used.

- The relationship between tables allows for the extraction of linked information.

- Allows users to be allocated different access rights to different parts of the database.

- Management information of a higher quality can be produced.

Disadvantages of database over flat file

- As all the data for a range of applications is held in one place there are greater security and confidentiality issues.

- Users will need to be trained to use the system.

Activity 1

Using Microsoft Access or another relational DBMS package, set up an application to store details of a CD collection. Use the following attributes (fields):

- CD Code

- Title

- Artist

- Label

- Number of Tracks

- Type of music (select your own categories)

You need to carry out the following stages:

- Design the table, selecting the appropriate data type, field length and properties for each field.

- Design a form for entering details of new albums.

- Add details of around 20 CDs.

- Design a report that will list all CDs held as well as a summary showing the total number of records and tracks for each category. You could also group the CDs by artist.

- Using QBE write queries to

  - List Title, Artist, Type of music and number of tracks in alphabetical order of Artist

  - List the CD Code and Title of all the CDs held by a particular Artist, whose name is entered at run time.

Use the Help facility of Access.

Activity 2

Using Microsoft Access or another relational DBMS package, set up an application for a veterinary surgery.

You need to carry out the following stages:

- Design the Pet and Owner tables, using the fields given in the chapter and adding some more of your own. Select the appropriate data type, field length and properties for each field. Remember the field Owner Code creates a link between the two tables.

- Set up a relationship between the two occurrences of the Owner Code field. (Ensure referential integrity).

- Design forms for data entry to each tables.

- Add details of about 5 owners.

- Add details of about 20 pets. You should not be able to enter an owner code in this table for an owner that you have not added to the owner table.

- Write queries to:

  - Output the number of each type of pet
  - Output the names and addresses, in alphabetical order, of all the owners of a pet whose Type is entered at run time.

Activity 3

For each of the scenarios described a database will require two tables. Name the tables and list appropriate attributes for each. Show how a foreign key as the link between the tables. Choose the attribute that is to be the primary key in each table with care, and justify your choices.

1. A conference has been organised for students of IT A level in a large city and students are coming from a number of schools. The organisers wish to store data about individual students and the schools involved.
2. The organisers of an arts festival wish to store data about venues where events are held the artists who are taking part (assume that an artist only performs at one venue).
3. Holiday company works with hotels in a number of resorts.
4. An art dealer sells the paintings and other works of art produced by a number of artists.

Activity 4

Why should a video shop use a database? Could a manual system be better than a database system? Both systems record details of members and who has hired which video. (This could easily be done in a paper based system by using a list of all videos and the name of the hirer written next to it.)

A manual system is cheaper, unlikely to break down and requires little training. However the computerised system will probably be better for the following reasons:

- Management information is automatically gathered, for example details of each hiring, financial details, how many times a customer has hired a video and how many times a video has been hired. These figures can be used in preparation of accounts or to analyse which videos are most popular.

- Better service to customers. Using a bar code reader to enter the video code and the member code is very quick. Queues at the counter will be shorter.

- Details of members and videos can be found and printed quickly.

- The names and addresses of members can be used for advertising purposes in a direct-mail shot. The database can be queried to come up with a list of people who haven't hired a video for six months and a letter written offering them a discount if they hire a film this week. The letter can be personalised using the mail-merge from a word-processing package.

- Similarly automatic reminders can be sent out to members who have not returned a video by the due date.

- The database can be extended to include the member's date of birth. The computer can be used to ensure that a member is old enough to hire, say, an 18 video.

Describe in a similar form to the example given above, the advantages of implementing a database system for:

a) a veterinary practice
b) a private leisure centre
c) a knitware business.

Summary

The term flat file is used to refer to a single file that is like a 2-dimensional tables.

○ The use of **flat file** systems can produce data duplication where the same data item is stored in two or more different files.

○ Unnecessary data duplication is known as **data redundancy**

○ Redundancy often leads to **data inconsistency** where the same item of data is stored differently in different places

A **database** is a store of data that can be linked together in different ways. The software package that allows a user to set up and access a database is called a **database management system** (DBMS).

A **relational database** is made up of:

○ **tables** that contain the data, each column representing one attribute (or field) of the data

○ **relationships** or links between the tables that are made by duplicating fields

○ **forms** that are used for data entry

○ **reports** that are used to present data.

○ **Queries** that are used to select data from one or more tables of the database

○ **macros** written to automate processes.

A **primary** key is a unique field or group of fields chosen to identify a record.

A **foreign key** is a field in a table that is identical to the primary key of another table and is used to provide a link between the two tables. The format of the foreign key should be identical to that of the primary key in the linked table.

Advantages of database over flat file include:

○ the reduction of data duplication, redundancy.and inconsistency

○ there is a centralised pool of data

○ data is independent from the applications

○ ad hoc (one off) and complex queries involving data from a number of linked tables can be made

○ different access rights to different parts of the database can be allocated.

Disadvantages of database over flat file

○ greater security and confidentiality issues arise from all the data for a range of applications being held in one place

○ users will need to be trained to use the system.

Organisation of data for effective retrieval questions

1. A company makes use of a computerised flat file information storage and retrieval system. The company is experiencing problems due to the use of this flat file system.

   a) Describe three benefits that the company would gain by using a relational database as opposed to a flat file system. *(6)*

2. A medium sized electrical shop is to be computerised using a relational database. Two tables are called PRODUCT and MANUFACTURER.

   a) State eight fields associated with the PRODUCT. *(4)*

   b) State two other tables which could be related to either or both of the original tables.. *(6)*

3. A video and games rental shop uses a database package to record membership details of customers who wish to borrow videos or games. The following details must be entered for each new member; full name, address, postcode and date of birth.

   a) Name and describe a validation check that could be used when data is first entered into the following fields for a new member. Your validation check must be different in each case.

   i)   Full_Name *(2)*
   ii) Postcode *(2)*
   iii) Date of Birth *(2)*

   b) The database package automatically generates a membership number. Explain why this is required. *(2)*

NEAB Jan 2001 Paper 2

Customer File

Surname	Forename	Street	Town	Postcode	Telephone no	Account No1	Account No2
Smith	Jayne	11 The Avenue	Chelmsford	CH6 1UT	01245 789423	12345	23567
Henry	James	2 The Street	Colchester	CO7 4PR	01206 654321	12456	34588
Chan	Daniel	3 High Street	Ipswich	IP3 4GH	01473 459876	23672	36834

Account File

Account No	Account Type	Interest Rate	Balance	Amount	Date	Statement Frequency	Date of last Statement
12345	High Interest	6.5 %	£10034.89	£120.45	12/03/99	Annual	01/01/99
23567	Cheque	1.5 %	£123.78	−£23.78	12/03/99	Monthly	01/03/99
36834	Saver	4.5 %	£1204.76	£156.87	13/03/99	Monthly	23/02/99
34588	Saver	4.5 %	£875.99	£17.99	13/03/99	Annual	20/03/98
12456	High Interest	6.5 %	£2001.76	£23.99	13/03/99	Annual	01/01/99
12345	High Interest	6.5 %	£10155.34	£56.73	14/03/99	Annual	01/01/99
36834	Saver	4.5 %	£1361.63	−£50.00	14/03/99	Monthly	23/02/99

4. A company is experiencing difficulties with its computerised flat file information storage and retrieval system. Describe three benefits that the company would gain by converting to a relational database system. *(6)*

NEAB Jan 2001 Paper 2

5. Give four advantages of using a relational database rather than a flat file system. *(4)*

NEAB June 2001 Paper 2

6. A building society keeps records of its customers accounts and transactions in a flat file information storage and retrieval system. Examples from the flat files are shown below.

a) The building society is experiencing problems due to use of this flat file system. Describe three reasons why these problems might occur. *(6)*

b) The building society decides to implement a relational database system. Draw up tables to show how data could be held. *(8)*

c) Explain how a query could be used with your tables to produce statements of transactions to be posted to customers. *(4)*

Adapted from NEAB 2000 Paper 2

7. Database Management Systems provide facilities to extract data from a stored database.

a) Name **two** common methods of setting up a query. *(2)*

b) State **one** advantage and **one** disadvantage of the two query methods stated in part (a). *(4)*

NEAB 1999 Paper 2

8. A college library uses a relational database management system to operate a membership and loans system. Staff and students can borrow as many books as they wish at any given time.

a) Name three database tables that you would expect to find in this system. In each case, identify the columns and keys required to enable this system to be maintained with minimum redundancy. *(6)*

b) Explain how the database tables named in part (a) would be linked. *(3)*

Adapted from NEAB 1997 Paper 2

9. The manager of a video hire shop uses a relational management system to operate the business. Separate database files hold details of customers, video films and loans. Customers can hire as many films as they wish.

a) For each of the files mentioned above identify the key fields and list other appropriate fields that would be required to enable this system to be maintained with minimum redundancy. *(6)*

b) Describe **three** advantages of using a relational database rather than a flat-file information storage system. *(6)*

NEAB 1996 Paper 2

10. College information system currently uses three sets of files for its student records, staff records and finance records, each being run separately. The system manager is keen to introduce a Database System (DBMS) based on a relational database, claiming that this will have major benefits for the college.

a) By the use of an appropriate example explain what is meant by he term 'relational database' 5 b. Describe three advantages of a DBMS approach in contrast with the use of independent files. *(6)*

b) Describe three advantages of a DBMS approach in contrast with the use of independent files. *(6)*

c) During the design of the system a decision is made to restrict the access of different users in different ways. Describe three different restrictions that may be imposed upon different users. *(6)*

NEAB 1995

11. A travel company has a number of holiday villas in Spain that they rent out to customers. At the moment, all the data relating to customers and villas is kept on paper that is stored in the company's office. There is a list of villas that homeowners rent out, a list of people looking to book a holiday and a booking file.

a) The travel company intends to change the manual system to a computerised database system. *(4)*

b) Explain whether a flat file or relational database would be appropriate and give reasons. *(6)*

c) Describe the tables that would be needed if a relational database were to be set. To do this list all the fields that there would be in each table, identify the primary key and show how links would be made between the tables. *(12)*

d) Suggest three different reports that could be produced from these tables *(6)*

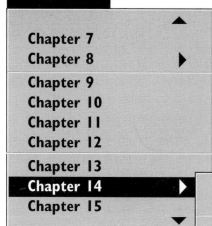

Types of software

Software means computer programs, thousands of instructions that are written to control the **hardware**; the physical components of the computer, such as printer, processor or disk drive. There are two main types of software:

○ systems software

○ applications software.

Systems software is the name given to the programs which help the user to control and make the best use of the computer hardware. The operating system and utility programs are systems software.

Applications software is a term used to describe programs that have been written to help the user carry out a specific task, such as paying wages, designing a newsletter or storing details of payments. Spreadsheets and word-processors are examples of applications software.

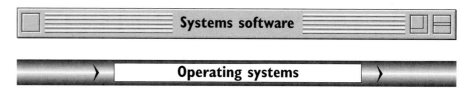

Systems software

Operating systems

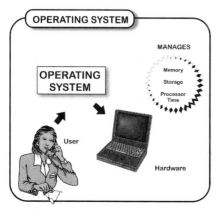

Figure 14.1 Functions of an operating system

An operating system is needed to run a computer. It controls and supervises the computer's hardware and supports all other software. It also provides an interface between the user and applications, and the computer's hardware. An operating system manages the computer's resources: memory, storage, processor time files and users.

Some specific tasks carried out by an operating system are:

○ memory management: allocating internal memory (RAM) – this is particularly important if the computer is running more than one program and the memory available may be relatively small

○ scheduling of programs (if more than one is running) according to pre-set priorities

○ transferring programs and data between disk and RAM

○ controlling input and output devices

○ 'booting up' the computer: carrying out the initial set up when the computer is switched on

- checking and controlling user access to prevent unauthorised access
- logging of errors.

An operating system may allow more than one program to run at the same time. With a microcomputer this is called **multi-tasking**. The programs are not actually running at the same time, as the operating system decides which program to run and switches from one to the other. This often happens so quickly that the user *thinks* that the programs are running at the same time. The operating system is managing the use of processor time to, make the very best use of it. When one program is tied up sending data to a printer the operating system can allocate the processor to another program. An operating system may allow more than one input device. The operating system checks which input devices are present and decides which should have priority.

The operating system manages the memory. It loads the program into RAM from disk and decides how much RAM to allocate to the program, how much to allocate to data and how much is needed for its own use. If there is not enough memory available, the operating system will give the user an error message. When several programs are running at the same time the operating system will need to keep track of where each one is stored and make sure that programs don't get mixed up.

An operating system also manages files stored on a disk. The user does not need to know where each file is physically stored, he can simply refer to the file by a name. It is the operating system that matches the file name to the actual data on disk.

Network operating systems are very complex. They will check usernames and passwords, restrict access to forbidden areas of the disk, manage the computer's memory and manage the printer queue.

Operating systems – Linux

Nearly all PCs use Microsoft's DOS operating system together with a Microsoft Windows graphical user interface. As well as giving Microsoft a virtual monopoly in this market, many users have commented that as PC hardware has developed, DOS has become inefficient, particularly for network servers.

Linux is an alternative operating system for the PC written by Linus Torvalds, available free over the Internet. At present, Linux is not yet suitable for desktop PCs, as there is little application software available. However, some users report that Linux is better for servers (computers controlling a network) than Windows.

In 1998, SBC Communications Inc. replaced 36 Windows 95 and Windows NT workstations at its Kansas City operations centre with

Figure 14.2 Linux logo

Linux workstations because they handled the results of a giant network monitoring system better. The graphics-intensive system caused the Windows 95 workstations 'to lock up on average every 4.2 minutes. The Windows NT workstations locked up every 2.58 minutes,' said Randy Kessell, a manager at the centre. The Linux workstations haven't had a problem.

Gary Nichols, manager of network administration at WaveTop, runs Linux on 30 of WaveTop's 45 servers for such tasks as e-mail, Web servers and the firewall. He worked out that he saved $30,000 in licence costs of Windows NT. 'I bought $100 worth of Linux CDs and books and got the same functionality,' he said.

Not surprisingly Linux has received financial support from Microsoft's rivals like Netscape, and Microsoft has acknowledged that Linux is one of the few threats on its horizon. However Linux will only threaten Windows when there is a large amount of application software available. At present there are no Linux equivalents for word processing, presentation graphics and spreadsheets. However in October 1998, Oracle released its Oracle 8 database for the Linux platform. Other software companies were also working on applications software for Linux such as SUSE's Linux Office Suite '99. Only time will tell if Microsoft has a serious rival.

Utility programs

Utility programs are programs that are tools to help you use your computer more effectively. Although utility programs are not part of an operating system, they are often provided with operating system software. Examples of utilities include:

- **file conversion**: converting files from one format to another. For example, a document file produced by Word will be stored with formatting codes that are specific to Word. For the document to be read by WordPerfect, a different word processing package with different formatting codes, a file conversion utility must be used.

- **file copying**

- **file compression** (sometimes referred to as zipping): many files, particularly graphics files, are very large. Data is a technique used to reduce the size of the file.

- **comparing the contents of two files or disks**: this program can be used to check that a file back up has been carried out successfully.

- **deleting files**

- **renaming files**

- **sorting data**

- **backing up a file**: making a copy of a file for security purposes

- **garbage collection**: this involves removing unwanted data and files from a disk and closing up any gaps left on the disk so that all the free space is together

> ## Configuring the system >

Today it is very common to want to be able to add additional input, output or storage devices to a computer. For example, you may wish to add a scanner, a joystick, a printer or a CD-writer. Whenever hardware is added, the computer must be set up to accept the new device. This is called configuring the system.

Printer drivers

The most common way of configuring hardware is using special software called a device driver. For example, different printers operate at different speeds, using slightly different sizes for lines and numbers of lines on a page. When you buy a printer, it is supplied with a program called a *printer driver*. Installing the printer driver formats the computer's output for the particular printer.

A printer driver provides communication between the operating system and the applications software. It translates formatting and highlighting information into a form that the printer can understand. It performs size control and enables error messages sent from the printer to be understandable to the operating system (such as 'out of paper').

A PC would need device drivers for a scanner or any new peripheral connected to the computer. Different screens will have different resolutions – disk drives will operate at different speeds. The driver handles communication between the operating system and the device.

> ## Applications software >

Applications software is usually broken down into two sections: general purpose or generic applications software and specific applications software.

Generic application software

The most common form of software is generic or general purpose application software. Generic application software is an applications package that is appropriate to many areas of day-to-day business operation. Many such programs are pre-installed on a computer when it is sold and can be used in many ways. You will use this type of software for your coursework. Generally, this software can be bought directly from the author or publisher, from a shop or by mail order through specialist computer magazines.

Many generic software packages can be used to create customised applications for a user. The use of macros, buttons and customised toolbars enable this to be done. Data can often be imported from one type of package to another. (See Chapter 15 for a more detailed discussion.)

The following are examples of generic applications software.

Word-processing packages are used to produce documents such as letters and reports. Text can be formatted in a variety of ways such as using different fonts, sizes, bold print and bullet points. Text can be edited using facilities such as copy and paste or find and replace. A modern word-processor has many features that were once only found in desktop publishing packages. A variety of objects such as symbols, graphics or other files can be inserted in a document.

Faster processing speeds and increased computer memory have allowed modern packages to include many extra features by mail-merge, macros, text-wrapping around imported graphics, e-mail, spelling and grammar checks and organising text in tables and columns.

Text written by one person can be edited by another. This book is written by two authors. When one of us completed a chapter we sent it to the other for checking and commenting as an e-mail attachment. The word-processing package has the facility to show all edited changes made by the reviewer. Deleted text is highlighted and added text displayed in another colour. Comments can be added as well. When the reviewed chapter is returned to the author he can accept or reject each change.

Database management packages are used for information storage and retrieval. These packages allow users to enter data and store it, to sort and search through the data and output information from the database in reports. More sophisticated packages will include wizards to help the user set up the database, import data from other packages and customise the database to hide the workings of the package from an inexperienced user.

Spreadsheet packages are used to store and manipulate tables of numerical data; automatically recalculating as the data is altered. They are ideal for storing, calculating and displaying financial information such as cash-flow forecasts, balance sheets and accounts. Today's spreadsheet users can set up several different types of graphs, perform many different mathematical functions and use macros, customisation and wizards.

Integrated packages are packages such as Works that contain a spreadsheet and a word-processor. Data can easily be transferred from one to another. Each part of the integrated package, such as the word-processor, typically has fewer functions than a full package of the same type. Integrated packages are usually easy to use, relatively cheap and offer sufficient functions to satisfy the occasional or inexperienced user.

Object Linking and Embedding (OLE)

Object linking and embedding provides a means of linking or sharing information between different programs such as spreadsheet and a word-processor. When using a word-processor to produce a document that is to be a report on sales a manager may wish to include extracts form a spreadsheet held in a spreadsheet together with some charts.

This can be done in two ways. The first uses an **embedded object** that is a selected part of the spreadsheet included as part of the document file. The document now actually contains all the data that makes up the object. The original spreadsheet can be modified or even deleted but the embedded object will remain unchanged. Using a number of embedded objects can result in a high storage requirement.

Using a **linked object** does not involve any storing of data in the document. Instead, a link is created to the spreadsheet file. If the data is changed in the spreadsheet the updates will be shown when the document is displayed. If the spreadsheet file were to be deleted then the link would be lost. Using linked objects allows reports to be kept up-to-date as data in other files is changed. Using linked files will minimise storage requirements as the spreadsheet data is only stored once.

Activity 1

Copy and complete the table below. For each type of software give the name of a package currently on the market. For each type of software, select the five features that you consider most important. Then write down a number of situations when it would be appropriate to use the type of software.

Software type	Name of package	Five features of software	Situations when software can be used
Word-processor			
Spreadsheet			
Database			
Integrated package			

Specific applications software

Specific applications software is designed to carry out a specific task, usually for a particular industry. It is of little use in other situations. Air traffic control software would be an example of specialised application software. A payroll application program would be another example. It is designed to be used exclusively for payroll activities and could not be used for any other tasks.

Specialised applications software is available for wide-ranging areas such as engineering and scientific work, which include a range of specialist design tools including specialist CAD packages see below and sophisticated mathematical software.

Simple graphics software such as *Paint*, which comes free with *Windows*, stores graphics files in bitmap form. This stores the colour of each pixel (dot) in the picture. As a result, bitmaps are large files and when they are resized, the image tends to be distorted.

Sophisticated graphics packages, like *CorelDraw*, store graphics files in vector graphic form. This means it stores pictures as a series of lines, arcs, text, etc. When storing a line, it will store the co-ordinates of the start and finish of the line, its width and colour. Not only does this save space, it means that the line can be moved, deleted or changed in colour or width without affecting the rest of the picture.

Gifs and jpg images are used on the Internet. They use compression techniques to store graphics files. This means that they are a fraction of the same picture in bitmap format. Graphics programs such as Paint Shop Pro and Microsoft Photo Editor can convert between different file formats.

Presentation software such as Microsoft PowerPoint has become increasingly common with the development of the Liquid Crystal Display (LCD) Projector. These projectors can project a computer display onto a large screen and are ideal for a presentation to an audience, usually replacing the old OHP (Overhead Projector).

Presentations in PowerPoint can include text and graphics, displays can be animated to attract the audience's attention and only display part of a page at a time. Sound files can be added for extra effect. Presentations can be stored on disk and edited for later use (see Chapter 17 for further discussion).

CAD (Computer-aided design) programs are used for designing, for example in engineering or architecture. Users can draw accurate straight lines and arcs of different types and thicknesses. By zooming in, designs can be produced more accurately. Designs can be produced in layers to show different information, for example one layer might show electrical wiring, another gas pipes and so on.

Network software

Web browsers such as Internet Explorer or Netscape Navigator allow users to access the Internet and view Web pages. You can store the addresses of 'favourite' pages and details of pages visited in a 'history' folder. Pages loaded previously are cached locally, i.e. stored on your hard drive to reduce loading time if the user decides to go back. A browser allows a user to go back to previously viewed pages.

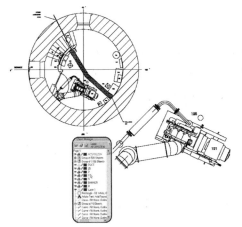

Figure 14.3 CAD

E-mail software such as Outlook Express allows users to send and receive e-mails, to store an 'address book' of e-mail addresses and to set up mailing lists of groups of recipients. Other documents such as graphics files can be attached to an e-mail. A reply button enables you to start a new e-mail to a correspondent without having to enter the e-mail address. An e-mail can be forwarded to another e-mail address.

Web editors such as FrontPage or Dream Weaver allow users to set up their own Web pages, including inserting images and adding links to other pages. Advanced editing software will set up frames within pages and eye-catching features such as hover-buttons and animations.

Bespoke software

Application software can be bought 'off the shelf' or can be written especially for a user. Specially written software is often called **bespoke** or **tailor-made** software. Bespoke software can either be developed in-house, by the programmers employed by the user's company, or by using an outside agency.

In-house software is designed to meet the exact needs of the user by personnel employed within the firm. If expertise is not available in-house it is possible to contract a software house to develop software to meet specific user needs.

Advantages of using pre-written software

Pre-written software is ready to use immediately and less expensive than bespoke software. The user can be confident that it has been fully tested. Some pre-written software can be customised to fit the individual user's need.

However bespoke software can precisely fit user requirements and may be the only way of solving the problem.

Case Study 1

Bespoke Software – BBC

The BBC needed to train staff to use their outside broadcast presentation equipment. However, the equipment was very expensive and in use nearly all the time. The BBC decided to use a simulation of the system for staff training.

As no such simulation software was available, a bespoke solution was the only option. The BBC commissioned Solos Interactive Limited to develop and produce the software which would simulate the outside broadcast equipment. Now training can take place without the need for expensive equipment.

Case Study 2

Bespoke Software – Caterpillar

Californian software giant i2 produces bespoke software for clients. They have developed real-time order management systems with customers, such as Siemens, K-Mart and Caterpillar. The software, which needs to be tailored to the demands of the customer, helps them manage their suppliers, plan orders and make production more efficient.

i2 has had a long-standing relationship with Caterpillar, who claimed that the bespoke system has enabled its engine products division to save $32m by reducing stock levels and improving assembly times by 38 per cent.

Summary

- Hardware means the physical parts of the computer. Software means the computer programs. There are two main types of software:

  - systems software – programs which help the computer run more smoothly. Utility programs and operating systems are types of systems software

  - applications software – programs that carry out a specific task, such as calculating staff wages: Word-processing, spreadsheets, database packages, graphics packages, web browsers and email are examples of applications software

- The tasks of an operating system include allocating internal memory (RAM), transferring programs and data between disk and RAM, checking user access and logging of errors.

- Examples of utilities include programs to delete, rename or copy files.

- Bespoke software is application software written especially for a user.

Software questions

1. You have installed a new piece of applications software onto a stand-alone PC?. You then find that the printer attached to the PC fails to produce what can be seen on screen in that package.

Explain clearly why this might happen. *(2)*

AQA ICT Module 2 Specimen Paper

2. When installing or configuring a particular word-processing package, the documentation states that the correct printer driver must also be installed. What is a printer driver, and why is it necessary? *(4)*

NEAB 1997 Paper 2

3. A new printer is supplied with printer driver files. The files are provided both on a floppy disk and on a CD-ROM. The CD-ROM also contains sound files for use with the printer.

Describe the functions of a printer driver. *(2)*

AQA ICT Module 2 May 2001

4. A small company is purchasing a new computer system and software. The new software includes an operating system, and generic package software which contains an application generator.

a) Give **three** tasks that are performed by an *operating system*. *(3)*

b) State **three** characteristics of *generic package software*. Illustrate your answer with **three** different examples of the type of packages that could be chosen by the company. (*The use of brand names will not gain credit.*) *(6)*

c) State **two** characteristics of an application generator (*see Chapter 15*) *(2)*

AQA ICT Module 2 May 2001

5. What is a utility program? Give an example. *(3)*

6. By giving examples, explain the difference between applications software and systems software. *(4)*

NEAB 1996 Paper 2

7. From the user's point of view, give three functions of an operating system. *(3)*

NEAB Specimen Paper 2

8. When using any applications software package on a network, the user is often unaware that an operating system is working 'behind the scenes', managing system resources. Give three of these resources and in each case briefly explain the role of the operating system in its management. *(6)*

NEAB 1997 Paper 2

9. State two editing facilities that are offered by word processing software. *(2)*

AQA ICT Module 2 January 2001

Capabilities and limitations of software

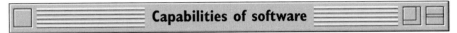
Capabilities of software

There are many features which are desirable in software and should be looked for when choosing a suitable package.

Links to other packages

(see data portability page 176)

This allows for the embedding or importing of data from other packages. Files can be saved in the format of another package. For example, Microsoft Word can open files from other software such as Works, WordPerfect and Write.

Files can also be saved in these formats.

Figure 15.1 Drop-down menu showing selection of file types that can be imported into Word 2000

Backwards compatibility

(See compatibility page 177)

Software must be able to read files set up in an earlier version of the program.

Search facilities

Such facilities allow the user to move quickly to the desired data. Search facilities are of course a feature of database packages, but the ability to search for words or phrases using **Edit**, **Find** is valuable in other packages such as word-processors or spreadsheets.

Edit, Find is a feature of most software.

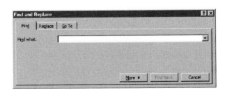

Figure 15.2 Edit, find

Macro capabilities

A sequence of keystrokes and menu choices can be saved as a macro. The macro can be activated by a single instruction or hot key, icon or button selection. Macros can speed up common operations, customise a package for a particular use and also allow a complex operation to be set up by an experienced user and used by less experienced ones.

Macros are used to customise the user environment or automate a task.

For example, in an office a macro can be used in a word-processing package which allows the user to select an icon from a tool bar (or an entry from a menu) that calls up the template for a particular type of document. The macro can move the cursor to different positions in the document where data needs to be entered.

Applications generators

An application generator allows a developer to specify the interface, the input and output required, and functions required by a user. The code is then automatically generated to produce a customised application. This facility allows a knowledgeable user to customise a generic software package without specialist programming knowledge. For example, it will typically allow data capture forms and menu systems to be produced without the need for an extensive programming code to be undertaken. Microsoft Access, for example, has a wizard that allows the user to set up easily data entry forms like the one shown below.

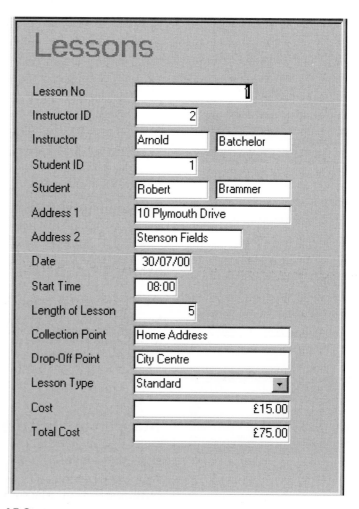

Figure 15.3 Application generators

Editing capabilities

Modern software packages allow the user to move around in a document or sheet using mouse or cursor keys, delete and insert data in the appropriate place with ease. Facilities to 'cut and paste' and 'search and replace' are standard. Some packages allow the editing to be carried out by another person; suggested changes are highlighted and the author can then accept or reject the changes as she wishes.

Ability to change or extend data and record structures

Flexible database packages allow for fields to be added or deleted when data is already recorded. Suitable safeguards are usually included to ensure that data is not lost unintentionally.

Short access times

Packages need to be able to retrieve data as quickly as possible. Speed of access can be a major factor in software selection.

Report generators

Report generators can take data from a database and turn it into a report. The format of the report and its wording can be defined by the user and customised to suit the application.

The report may be in graph, tabular or text form. Data can be added to give totals and sub-totals as required. The advantage of this sort of software is that once the report has been specified with this year's figures (for example), when next year's figures are available, the report can be prepared automatically.

Microsoft Access has a report wizard that allows the user to set up easily reports like the one shown.

Pass-It *Instructor's Timetable Report*   30 July 2000

Instructor ID		1	Doug	Jones

Start Time	Name			Length of Lesson
08:00	Robert	Brammer		1
09:00	Steven	Jenkins		2
11:00	Mary	Trueman		1
12:00	Victoria	Spencer		3

Instructor ID		2	Arnold	Batchelor

Start Time	Name			Length of Lesson
12:00	David	Windsor		1

Figure 15.4 Report

Data portability

The ability to transfer data to or from another package or hardware platform is a feature that is increasingly becoming an important requirement for users. This means data does not have to be typed in again, which would waste time and could lead to errors.

Data is said to be **portable** if it can be transferred from one application to another in electronic form. (Portability has a specialist meaning here – it doesn't mean 'you can put a floppy disk in your pocket and carry it around'!) It is very common to need to transfer data produced in one package to another package. Different packages have different functions, many of which are used by businesses today.

A user may have different packages, or versions of packages, available at home and at work and will need to be able to transfer data. The growth in the use of networks has increased the need for portable data files. Portability ensures that documents produced on one package can be accessed by other similar packages, or by the same package on different hardware platforms.

For example, a freelance journalist carries out much of his work at home using Microsoft Word on a PC. One of the magazines he works for has a network of Apple Macs on which they run Microsoft Word as their word-processor. The second magazine uses WordPerfect on PCs. The journalist needs to be able to transfer documents produced at home to either place of work.

Consider a sales manager who is writing a report on the performance of her sales representatives during the past year. She is producing the report using a word-processing package. Details of sales throughout the year are maintained on a spreadsheet package that has graphing capabilities. The sales manager would like to include graphics and tables into her report. Ideally this data could be **imported** from the spreadsheet into the word-processor.

Users buying a new package will want to be able to import files from their previous package. Portability is an important marketing feature for companies producing new software. Early microcomputers had no common standard for storing data. As a result it was very difficult to transfer data between computers made by different manufacturers. Now manufacturers have standardised to PC format, making it easy to transfer for data between different computers – for example data can be quickly transferred from a palmtop to a PC.

Portability and Windows

Windows offers portability – the ability to take data from one program to another. By using **Edit** and **Copy**, you can copy data from one program into the Clipboard. You can then use **Edit**, **Paste** in another program.

Windows programs usually allow you to import and export data from and to another program. Many Windows programs are published by the same company, Microsoft. For example Word, Publisher, Excel and Access. You can import into Word or Access from Excel into Publisher from Word and into Excel or Word from Access.

Many software suppliers provide filters that allow documents developed in other companies' packages to be converted into a suitable format for input. Word 2000 can import from WordPerfect 5.*, or WordPerfect 6.*. A WordPerfect user could buy Word 2000 and not lose previous work.

There are several different types of picture format, for example gif, bmp and jpg. Word 2000 can import from at least 20 different formats of picture.

Upgradability of software

Every few years a new version of the same program is produced. New versions called upgrades take advantage of increased speed, memory and processing power of later computers to offer new functions.

To distinguish different versions, they have a number such as Microsoft Internet Explorer 6 or Microsoft Word 2000.

Microsoft Word 95 added an automatic spell check to the previous version, underlining a mistake as you type. The next version, Microsoft Word 97 added the function to save a file as a web page. Word could be used as a web editor. The next version, Microsoft Word 2000 added the ability to edit and send e-mails within Word. These functions reflected the increased importance of the Internet and e-mail.

Major upgrades are usually numbered as '.0' versions (e.g. 4.0, 5.0). Minor changes are numbered as .1 or .2 or even .01 versions.

Compatibility

When upgrading your software, for example, from Microsoft Office 97 to Office 2000, you will still want to be able to load your old Office 97 files. It is essential that software upgrades are backwards compatible. For example, Microsoft Access 2000 can convert Access 97 files to Access 2000 format and open them.

It is often not true the other way round. For example, files saved in Access 2000 cannot be opened in Access 97. If you want to use an Access 2000 file in Access 97, it is necessary to use a utility in Access 2000 to convert the file to Access 97 format.

Access 2000 has a utility to convert to Access 97 format.

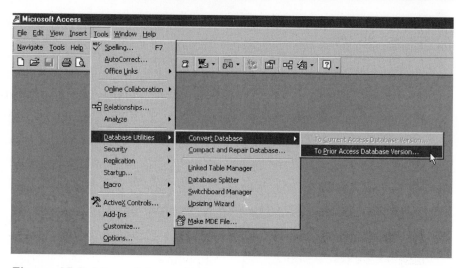

Figure 15.5 Microsoft Access 2000 can convert a database file to work in a previous version of the software

Software upgrades should be sufficiently similar to the previous version so that users are immediately comfortable with the new version. The increased processing power of recent computers means that upgraded software can be even more user-friendly.

When a new program is brought out it is important that:

● it has additional new features which make it attractive to the buyer

● a user can load work done in the previous version of the program (backwards compatibility)

● it looks similar enough to the old version so that users can start straight away.

Reasons not to upgrade

New versions of software may have more features and be more efficient than earlier versions. However users may not want to upgrade to the latest software for some of the following reasons:

● the current hardware specification may not be suitable for the new software, for example the processor speed may be too slow or the memory inadequate

● the cost of the new software may be prohibitive

● the new features offered may not be required

● training may be needed for the new software

● installation of the new software may mean that the equipment is out of action for a time.

Case Study 1

Portability – Cranmer Estates

Cranmer Estates is a firm of estate agents with around ten branches in the north of England. Until recently they have been using old PCs, which did not run Windows, and a DOS version of the word-processing program WordPerfect 4.1 to produce details of houses for sale and to type letters.

Owner Derek Cranmer decided to invest in new equipment because he wanted to store details of houses for sale in a database, use a spreadsheet to keep accounts and to be able to send house details from one branch to another by e-mail. He decided to buy new Windows PCs with e-mail facilities running a word-processing package, a database package and a spreadsheet package.

Derek considered many packages such as Microsoft Office Professional, Corel Office Professional (which included the latest version of WordPerfect) and Lotus Smart Suite. His priorities in deciding were:

1. Portability. It was essential to be able to import old WordPerfect files as he did not want his staff to have to type data in again.

2. Links to other packages. He also wanted to be able to take data about houses from the database into the word-processor.

3. Training. He wanted to use a package where training was available. Derek found out that his local college ran courses in Microsoft Word, Excel and Access – all part of Microsoft Office.

The need for training persuaded Derek to choose Microsoft Office, which could also import from WordPerfect. After a pilot project at one branch was successful, Derek introduced the new system into all his branches.

Reliability of software

We all want software that is bug-free and will not crash while we are using it. However software bugs are not uncommon.

To produce 100 per cent bug-free software is a near impossibility. A program will contain hundreds of thousands of lines of program code with millions of different pathways through it and this requires very rigorous testing.

In the software industry there are strong commercial pressures to produce your software before the competition and to keep development costs down. Often software is published without being fully tested.

Bugs often only come to light when the software is used on a particular specification of PC or in an unplanned situation. Producers may need to distribute program 'patches', which repair errors that have been discovered. Upgraded .1 or .2 versions of software may be distributed free or at considerable discount to existing users of the .0 version.

Testing is a vital part of software development. Programs should be tested by colleagues who have not been involved with the actual production themselves. Then they will be given to selected users to try out in real situations and report bugs before the official release date. This means that they can be tested on different hardware platforms, with different memory sizes and processor speeds that may operate differently.

Summary

○ Today most packages have links to other packages allowing them to share the same data. Data is said to be portable if it can be transferred from one application to another. Windows software offers portability.

○ New versions of software are similar to the previous version so that users can use the new version straight away. The new version should be able to load work saved in the previous version.

○ New software must be tested thoroughly before being released commercially. However time and financial pressures, and the number of different paths and platforms, means that software may still have bugs when it is sold.

Capabilities and limitations of software questions

1. Two users are using two different versions of the same word-processing package. When user A sends a file on disk to user B there are no problems in reading the file. However, when files are transferred the other way the transfer is not successful. Explain why this may happen and how to overcome this problem. *(4)*

NEAB Specimen Paper 2

2. Why does commercially available software not always function correctly when installed onto a computer system? *(2)*

AQA ICT Module 2 Specimen Paper

3. A spreadsheet package has macro capabilities.

a)   Describe what is meant by the term macro. *(2)*

b)   Give two examples of situations where the use of macros would be appropriate. *(2)*

AQA ICT Module 2 January 2001

4. A freelance reporter who regularly contributes articles to various newspapers and magazines is considering which word-processing package she should purchase. A friend has said that 'most modern application packages enable users to produce files which are portable'.

a)   Explain what portability means in this context. *(4)*

b)   Explain why portability is important. *(4)*

5. There is now a wide range of software tools available to increase the productivity of the end-user. Two such software tools are Application Generators and Report Generators.

a)   Explain what is meant by an Application Generator. *(2)*

b)   Explain what is meant by a Report Generator. *(2)*

c)   Give an example of when it might be sensible to use each one. *(2)*

AQA ICT Module 2 Specimen Paper

6. Users may encounter problems when software manufacturers upgrade a software package. With reference to specific examples describe two such problems. *(4)*

7. Articles in the media referring to computer software which fails to work properly are commonplace. Discuss the difficulties facing software companies when testing and implementing complex software, and the measures that software providers could take to minimise these problems. *(6)*

1996 Paper 2

Modes of operation

All computers work on a basis of

Input   Process   Output

Different computer systems require different modes of operation that determine how the computer is used. Sometimes it is appropriate or necessary to process each occurrence of data as it presents itself, whilst in other situations it is more appropriate, efficient and cheaper to collect a large amount of data before processing it all together.

For example, if you go to a bank's Automatic Teller Machine (ATM or 'hole in the wall') to find out the current balance of your account, you want an immediate response. You would not be happy to wait for 50 other people to make a similar request before yours was dealt with! If the computer system was one controlling a chemical process, by inputting data on factors such as temperature through sensors and making necessary changes such as turning on a heater, anything but an immediate response could prove disastrous. On the other hand, the most efficient way to enter the responses to a survey involving several hundred people is to collect them together and read them automatically, by using an input device such as OMR or OCR (see Chapter 11).

A number of factors will determine the mode of operation for a particular system. These include hardware availability, the volume of data, the required response time as well as the nature of the system. The four modes of operation to be considered are:

- real time

- batch

- transaction

- interactive (or pseudo real time)

Before investigating these in more detail it is necessary to define some of the terms that will be used.

Master file	A **master file** is a main file of records relating to a system. For example, a banking system would have a master file of account details with one record for each account; a stock control system for a retailing business would have a stock master file where each record held details relating to one item of stock.
Transaction	A **transaction** is a single change to a record held in a master file.
Transaction file	A **transaction file** is made up of transaction records and is used to update a master file with changes.
Off-line processing	**Off-line** processing occurs away from the main computer. Data can be recorded onto a magnetic device such as disk using a separate device.
On-line processing	**On-line** processing occurs under the control of the computer.

Real time

A system running in the real time mode of operation is one that can react fast enough to influence events outside the computer system. An air traffic control system is as example of a system operating in real time mode. Most systems operating in realtime mode use a processor that is dedicated to that specific system.

For example, a computer controlled chemical process uses sensors to input the temperature of the substances into the computer. The computer's processor processes this information, controlling outputs that turn on heaters and open valves when needed. This is an example of real time processing. The computer is operating all the time, receiving data, processing it and outputting information in time to influence events. There are many examples of control situations when real time processing is used. These range from very large systems such as the system that controls the operation of a nuclear reactor to small embedded microprocessor systems that control the functioning of devices such as washing machines, a service station petrol pump or a burglar alarm system.

The Tropical House at Kew Gardens provides an environment that enables plants from the tropics that would not survive in our climate to flourish. To maintain the correct environment, sensors input such things as temperature and humidity and the computer system enables actuators to switch heaters on or off, open or close windows, or turn water sprayers on or off as appropriate.

The concept of feedback is essential to real time processing. Figure 16.1 shows what is involved in a system that monitors temperature. The computer has to process the data input from the temperature sensor in time to turn on the fan or the heater so that the system is maintained within allowable temperatures.

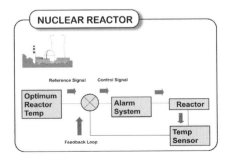

Figure 16.1 A feedback loop for monitoring the temperature

Batch processing

Batch mode is another common method of computer processing used when there are large numbers of similar transactions. All the data to be input is collected together before being processed in a single operation.

Early computer systems, using large mainframe computers, all operated in batch mode. Many of these systems have been replaced by real time processing.

However, there are still a number of tasks that are best carried out in batch mode. Batch processing is the most appropriate and efficient mode to use when a large number of records in a master file need to be updated.

In a typical batch processing system:

1. All the transaction are batched together in an off-line transaction file.

2. This transaction file is sorted into the same order as the records in the master file.

3. The record from the transaction file is merged with the corresponding record of the master file.

4. The updated record is stored in a new master file.

5. This is repeated for every record in the transaction file.

The old master file is called the **father file** and the new version the **son**. When the son is in turn used to create a new version it becomes the father and the old father becomes the **grandfather**. A number of generations can be kept for back-up purposes. As long as the relevant transaction files are also kept the current master file can be recreated.

Example of batch processing – a billing system

An example of a typical batch processing operation is an electricity billing system. The computer already has much of the data required to calculate the bill stored on disk in the master file. The data fields stored will include:

-   customer number – the unique code used to identify the customer
-   name of customer
-   address
-   last meter reading – to calculate the number of units used and the charge
-   amount of electricity used in last four quarters – for checking that this quarter's usage is sensible
-   special instructions for meter reader – for example, 'the meter round back of the house'

1. The meter reader writes the meter reading on the card. When the card is produced, high and low estimates for the reading are printed so that the reader can check that the value he enters is sensible. It is important that no transcription error is made. At the end of the day the reader hands in the completed cards at the office (see Chapter 11 for other methods of capturing the data).

2. The data from the cards is prepared off-line for updating the master file and a transaction file is created. Each transaction record will need two items of data: the customer number, so that the correct record in the master file can be identified, and the current meter reading.

3. The cards are collected together into batches for about 100 houses. A control total, containing the number of cards and the sum of all the meter readings (calculated manually), is added. This is used as a check (see Chapter 12).

4. The data is entered into a disk file using a key-to-disk system. A verification check is usually carried out at this stage: a second person types in the data and it is checked against the first entry.

5. Once the data from all the cards has been entered the transaction file is complete. It is then read into a data validation program that checks that each customer number is valid by recalculating and checking the check digit, and that each meter reading is within an allowable range. The number and sum of the readings are calculated and compared with the batch totals. Any errors detected in the validation program are printed in a report for investigation and correction.

6. All correct transactions are copied into a valid transactions file. The transaction records in this file are not in any order – they have to be sorted into the same order as the master file. When this has been done, the two files can be read through together, record by record, the customer number being used to match transactions with master file records.

7. Each time a match is found, the details of the bill can be calculated. The bill will be printed on special, pre-printed stationary using high speed printers. A record is created for a new version of the master file which will contain data updated from the transaction record. This new version of the master file will be used in three months' time when the next reading of the meter is made.

When the new version of the master file is created, the old version is still kept as it provides back-up. If the new version were lost or corrupted, it could be recreated by running the update program again using the old master file and the transaction file.

Using batch processing for this billing system means that details of only one customer are processed at one time – the whole file does not have to be loaded in – so that the computer does not need a large amount of memory. Files can be stored on magnetic tape or magnetic disk. If magnetic tape is used the computer would have to have three tape drives – one for the transaction file, one for the father master file and one for the son master file.

Very large volumes of data can be processed efficiently in one run by using batch processing. Each input file is read through just once from start to finish. Once the program is running there is no human intervention. All the data has to be entered at the start and no more can be entered once the system is running. This means that information can be out-of-date and new data can only be entered with the next batch. You cannot perform an immediate search for information as response time for a specific query is slow.

Batch processing is suitable for payroll or billing systems, which are run once a week, once a month or once a quarter. It is not suitable for a system that needs an immediate response, such as an enquiry system.

Activity I

Outline the stages involved in a batch processing payroll system for a large organisation that inputs the number of hours worked by employees and produces pay slips.

Figure 16.2 Building Society passbook

Transaction processing

Although there are situations where batch processing is still the most efficient mode of operation, many systems that used to operate in batch mode now allow each transaction to be processed as it occurs.

Transaction processing is a mode of operation where data for each transaction is entered at source and processed immediately. In a building society customers can be issued with passbooks. If they wish to make a deposit customers can take their passbook, together with a completed deposit transaction form, to a till at any branch office. The building society clerk will enter the details of the transaction (account number and amount of deposit) using the keyboard of a networked computer (DDE). The account details record in the master file is updated to reflect the transaction, the old values are over-written. The passbook is placed into a printing device and the details of the transaction and the new balance are printed.

Unlike batch processing the master file is updated with the data from the transaction immediately. Only one record of the master file is accessed, so the file must be organised in a way that allows each record to be located directly and independently. The great advantage of transaction processing is that the master file is always kept up-to-date. However, master file back-up becomes more of a problem as there is no automatic production of back-up files such as when batch processing mode is used.

Many systems use a combination of processing in batch and transaction modes. The building society needs to produce annual account statements for members who may need the details for tax purposes. A printed summary is needed for each account. This is most efficiently produced automatically, in batch mode. It would be a long task if transaction processing were used, as a clerk would need to enter each account number in turn to initiate each transaction.

Interactive (pseudo real time) processing

A particular form of transaction processing is interactive processing. A characteristic of real time processing is that the response time is sufficiently fast to influence critical events external to the system. Interactive processing requires a quick response in human rather than in machine terms. A pseudo real time system accepts transactions from outside sources and processes one transaction before another one is accepted. A typical example of an interactive system is a theatre booking system where one booking must be completed before another one is started. In this way there is no danger of the same seat being booked out to two different people. Much home computer use involves interactive processing.

Activity 2

For each of the systems below identify and *justify* the appropriate operating system processing mode:

Airline reservation system
A user sitting at a terminal will type in details of the customer's request, and details of suitable flights will be displayed. A booking can then be made and will be processed immediately.

A nuclear reactor
A computer is used to monitor the temperature stability of the nuclear reactor. If this temperature becomes critical then an alarm system is activated.

Play-a-Toy Ltd
Playa-a-Toy is a large manufacturing company with around 6,000 employees. On the last working day of every month the computer system automatically transfers wages to the employees' bank accounts and produces individualised pay slips.

British Telecom
Every three months British Telecom calculates a bill for each of its thousands of customers and prints the details on to pre-printed paper for posting.

The home user
Mary has a computer at home with a modem connection and access to the Internet. On occasions she uses a browser to surf the Internet and down-load interesting articles.

continued

Activity 2 *continued*

Retailer

A local retail store has recently upgraded its premises. As part of the renovations a new set of automatic doors and an advanced air-conditioning system has been installed. The doors allow customers to move in and out of the shop freely and the air-conditioning system maintains a constant room temperature (the temperature is adjustable).

Examination marking

When students enter multiple choice examinations they fill in an OMR sheet with their choices marked in pencil. The sheets from every school are sent to the examining board, who read the data into a computer system that calculates the results. Individual student marks are summarised on a list for each school.

HSBC Bank mortgages

When a customer requests a mortgage, an advisor sits down with them in front of a computer. A program guides the advisor through a series of question for the customer and the answers are entered. At the end the program produces details of any possible mortgage offer.

American Express

Each month a bill is produced for each customer that lists the details of all transactions made in the last four weeks together with details of any payments made.

Zap 'Em

Zap 'Em is an arcade game in which buttons and joysticks are used to position a Zapper Gun to shoot down deadly aliens. These aliens are moving around the screen demolishing innocent Lemmings. As soon as a user hits a target, an alien is removed.

Case Study

Howse, Hulme and Byer, Estate Agents

Howse, Hulme and Byer is a chain of estate agents with 50 branches around the country. The current data processing system, which has been in place for a number of years' is based on batch processing.

The main categories of transactions that take place at the branch offices are:

- details of new potential purchasers

- details of a new house for sale

- details of a sale.

The current system

A branch employee writes details of a transaction on to a pre-printed form. These forms are collected and, three times a week, are sent to the head office where they are put together with forms from all the other branches. A data entry clerk keys in the transaction data off-line using a key-to-disk system. The data is entered again by a second clerk so that any mistakes made by the first clerk are highlighted.

As data from all the forms is keyed in it is stored on disk in a transaction file. This file is then inputted into a validation program that produces a valid transactions file as well as an error report that contains details of mistakes in the transactions. The reports are sent back to the branch offices so that the transactions can be corrected and re-submitted.

The file of valid transactions is then sorted into the same order as the master file and used to update the old master file by creating a new version that includes the changes resulting from the transactions.

An updated list of houses available, ordered by area and price, is sent to each branch. Lists of any newly available houses that meet their requirements are sent to prospective buyers.

Howse, Hulme and Byer are considering changing from a batch system to an interactive system. To do this, computers will need to be installed in each branch office. These will be linked to a central computer. Transactions will be processed as they occur, the details being keyed in at the branch office.

The new system would bring a number of advantages. As the transactions are entered into the system as they occur, the information available is more up-to-date; details of new houses are available in the branches as soon as they are received. The details of all houses for sale are available in all branches as soon as the transaction is complete. The confirmation of a sale is immediately recorded, thus preventing potential purchasers being shown the details of a house that is no longer available.

However, there are some issues that need to be considered before the system is updated. Obviously, there will be hardware and software cost implications. At present, all data entry is performed by trained staff in the head office; the new system will demand extra skills of branch employees; training will be needed in the use of the new system. There will be a greater security risk as computer records, at present only available at the head office, will be available at all the branches. Back-up files, which are produced automatically in a batch processing system, will require more complex organisation.

- Draw up a table of the advantages and disadvantages of moving to a new, interactive system. Refer to the text and add some ideas of your own.

Activity

A chain of retailers has 120 shops throughout the United Kingdom. At present Kimball tags are used to record sales. These are small hole-punched cards. Whenever a sale is made, the card is removed from the shoe box and collected in a bag. At the end of the day the cards are sent to the head office where the data from the cards is used to update the stock level records and re-orders are made automatically.

The management have decided to replace the current system with an interactive one.

○ What is an interactive system?

○ What type of system is the current one?

○ What method of data capture would you suggest for the new system?

○ Describe the stages of processing that currently take place at the head office.

○ Describe the advantages that would come from the new system.

Data characteristics

All data that is stored and processed by a computer is in binary form. Different binary codes are used to represent different types of data. The smallest unit of storage is a bit; a bit can be in one of two states: one state is used to represent a 0, the other state a 1. By building up combinations of bits, different codes can be stored. Two bits can store any one of four different codes: 00; 01;10 or 11. Three bits can store one of eight codes:

000	100
001	101
010	110
011	111

All data stored in a computer, whether text numbers, images or even sounds, are stored as a binary code of 0s and 1s. Programs are also stored in binary coded form.

Text

In many systems data is stored in text form. The word-processor is the most common software package that processes text data. Each character that can be used is assigned a binary code. Standard codes have been agreed; the most widely used is **ASCII** (American Standard

Code for Information Interchange). A character coded in ASCII is made up of 8 bits. The 8th bit acts as a parity bit to help ensure that any corruption of data is detected (see Chapter 12). The other seven bits can produce 128 different combinations, so 128 different characters can be represented.

The ASCII character set is made up of:

- upper case letters, A – Z

- lower case letters, a – z (note upper and lower case have different codes)

- digits 0 – 9

- special characters such as space, ; & %

- control characters such as escape, return, backspace, line feed, escape. These characters have binary codes between 0000000 and 0011111 (0 and 31 in decimal)

The alphabetic characters together with digits are often referred to as **alphanumeric** characters.

The ASCII coding system does not provide enough possible codes to include all the characters that may be required by the languages of the world. An extended system, called **Unicode**, using 16 bits per character, has now been developed.

Text is stored in lines; the CR/LF (carriage return, line feed) character indicates the end of a line. (The terms carriage return, and line feed comes from the old typewriter usage when the end of line is reached on a manual typewriter, a lever has to be pulled to return the roller holding the paper to the start of a line and rotate the roller for a new line).

Word-processing software packages include the facility to store documents in ASCII (often referred to simply as text) format. This allows a document to be transferred from one package to another. However, documents that are only used within the package will contain other codes that denote formatting, such as margins and the use of fonts. Each word-processing package has its own formatting, codes that are stored together with the character codes.

Graphics

Increasingly, computers are used to manipulate, store and display non-textual images. These graphics, or pictures, also have to be stored in binary coded form. There are two main ways in which images are stored: either as **bit-mapped** or **vector graphics**.

Bit-mapped graphics

A bit-map is the binary stored data representing an image. A picture is broken up into thousands of tiny squares called **pixels**. The more pixels that are used per square centimetre, the greater the resolution of the image and the better the image looks.

Each pixel is allocated a number of bits in the bit-map to represent its colour. The more bits allocated to each pixel the greater the choice of possible colours, but the amount of memory required to store an image will also be increased.

If only two colours, black and white, are used then just one bit is needed to represent each pixel. Figure 16.4 shows a simple black and white image built up in pixels and the bit-map that would be used to represent it.

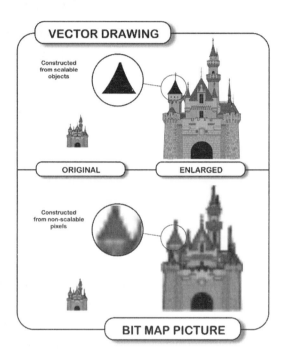

Figure 16.4 Bit-map and vector

As processors have become faster both the main memory (RAM) and the backing storage capacity have increased hugely. Modern computers are able to store and process complex images of high resolution and many colours.

Bit-mapped graphics can be created by using a drawing package where individual pixels can be set or lines 'drawn' using some kind of pointing device to modify an image displayed on a screen. An image can also be input using a scanner or a digital camera. A package can then be used to modify the image. For detailed changes the setting of individual pixels can be modified.

There are a number of standard formats that are used for storing graphical data. These are necessary, in the same way that ASCII is necessary for text storage, to allow graphical data to be transferred between different packages. An image developed in a painting package may then be used in a DTP package. One common format is **Tagged Image File Format (TIFF)**.

Data compression techniques are used to minimise the amount of storage space needed for graphical images. The **Joint Photographic Expert Group (JPEG)** has defined standards for graphical image compression. JPEG is now a commonly used format.

Figure 16.5 Bit-mapped image made up of pixels

Problems associated with bit mapped graphics

- Bit-mapped graphics can be difficult to edit.

- Image quality can be lost if enlargement takes place.

- Distortion can occur if the image is transferred to a computer whose screen has a different resolution.

- Large storage space is required to store the attributes of every pixel.

Computer animation is the creation of apparent movement by the display of a sequence of graphical images, each one differing very slightly from the previous one.

Vector, or object-oriented graphics

For applications such as CAD (Computer–Aided Design), where high precision is required, bit-mapped graphics are not appropriate. With object-oriented graphics the image is stored in terms of geometric data. For example, a circle is defined by its centre, its radius and its colour. Object-oriented graphics enable the user to manipulate objects as entire units for example to change the length of a line or enlarge a circle whereas bit-mapped graphics require repainting individual dots in the line or circle. Because objects are described mathematically, object-oriented graphics can also be layered, rotated and magnified relatively easily.

Case Study 2

The Trinity House Lighthouse Service

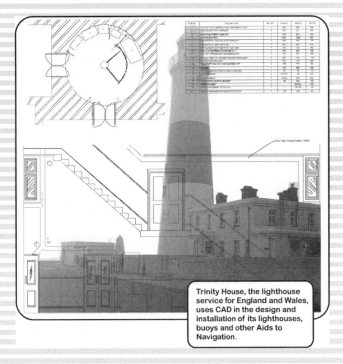

Trinity House, the lighthouse service for England and Wales, uses CAD in the design and installation of its lighthouses, buoys and other Aids to Navigation.

Trinity House provides nearly 600 aids to navigation sites such as storm-lashed lighthouses and buoys. Project Teams consisting of specialist engineers are responsible for projects from initial design through to completion.

A CAD program is used to generate detail and assemble drawings that are used in the manufacture, construction and installation of navigational equipment. The CAD program uses object-oriented graphics. This enables the engineer to represent the various components and services as objects that can be copied or adapted to serve different applications. Layering can be used to differentiate services such as water supplies, electrical cabling and control systems.

Input is normally via a digitiser and tablet with the primary output device usually being a pen plotter or A3 laser printer.

○ Why is a bit-mapped graphics package not appropriate for use by the Trinity House Lighthouse Service?

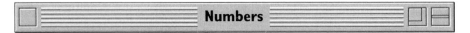

Numbers

Computers store numbers, as all other data, in binary coded form. There are three main ways of coding the numbers that we use.

Integers are whole numbers. When integers are stored in a computer, the number of bits assigned to the code determines the range of numbers that can be stored. The coding can be designed to store negative as well as positive integers. Integer arithmetic provides fast and accurate results – problems only occur if a calculation results in an integer that is too large to store in the number of bits assigned to the code.

A **real** number is a number that can have a fractional part. Real numbers are stored in a format known as **floating point representation**. Real numbers can rarely be stored exactly in the number of bits assigned to store the number thus floating point representation involves some loss of accuracy. The space allocated to store a real number, the greater the range of numbers that can be stored in the same way as intergers, but the accuracy of the representation of the number is also increased. Performing calculations on floating point numbers is a more complex operation than doing so with intergers and therefore is slower.

In systems where fractional values are needed but where accuracy is very important, then a third form of coding can be used. There are a number of applications where numbers that represent currency values are stored in a special format.

Sound

Sound travels in waves and is therefore analogue in form. To be stored into a computer the analogue signal must be converted into digital form. The wave is sampled through a microphone at regular intervals and an analogue to digital converter measures the height of the wave at that point and store it as a binary code. The more frequent the sampling, the more accurate the representation; obviously the amount of storage space required will increase as the sampling rate of the sound increases. When the sound is output the digital representation is converted back to analogue form and the signal output through a loud speaker. How closely the sound resembles the original wave will depend upon the sampling frequency.

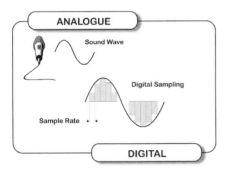

Figure 16.7 Analogue in, digital out

Summary

A computer system can use one of four modes of operation shown in the table below.. The appropriate choice of mode depends upon the nature of the application.

Processing mode	Description	Examples of use
Real time	Reacts fast enough to influence events outside the computer system.	Control systems such as a computer-controlled greenhouse or a guided missile system.
Batch	All the data to be input is collected together before being processed in a single operation. Used when volume of data to be processed at one time is large.	Invoicing systems; payroll systems.
Transaction	Data for each transaction is entered at source and processed immediately. The volume of data to be processed at one time is low.	On-line order processing systems.
Interactive (pseudo real time)	Transactions are accepted from outside sources and transaction is processed before another one is accepted.	Booking systems.

- All data is stored in coded binary form. Four categories of data are text, graphics, numbers and sound.

- **Text** data is stored in coded form where each code represents a single character. A commonly used code is ASCII, where each character is represented by 8 binary digits.

- **Picture** (or **graphical**) data is stored in either **bit-mapped** or **object** form. A bit mapped image stores a code for each pixel, the tiny squares of which such an image is made up. This code represents the attributes such as colour of the pixel.

  Object graphics store details of an image as a mathematical equation.

- **Numbers** can be stored in integer, floating point or currency format.

  Integers are whole numbers that can be processed fast with accuracy. The size of the number that can be stored is limited by the number of bits assigned to the code.

  Floating point representation is used to store real numbers where there is a fractional part. Very large and very small numbers can be stored but full accuracy cannot be achieved.

- Sound is stored in digitised form where the natural wave patterns are converted into binary codes using an analogue to digital converter. The quality of the sound stored depends upon the sampling rate.

Modes of operation questions

1. A bank's computer supports real time enquiries in the day, but at night works in batch processing mode. Suggest three tasks that the computer would perform in batch processing mode. (3)

2. A chain of estate agents has 80 branches. Daily transactions relating to house sales, purchases and enquiries are processed using a batch system based on a mainframe computer at head office.

   a) Outline the flow of data through such a batch processing system. *(4)*

   b) The company is considering changing from the batch system to an interactive system. Describe the advantages and disadvantages of moving to an interactive system. *(4)*

   NEAB 1996 Paper 2

3. A company is about to replace its old batch processing system for the preparation of customer accounts, by a real time system.

   a) Give two distinct advantages, which could be expected as a result of this change over.

   b) Suggest two problems that are likely to be encountered during the changeover period. *(4)*

4. List the circumstances when batch processing is more appropriate than real time processing. *(4)*

5. A computer system can be described as being a 'pseudo realtime system'.

   a) State clearly what is meant by pseudo real time. *(2)*

   b) Give a situation where pseudo real time is essential, stating a reason why it is needed. *(2)*

   NEAB 1998 Paper 2

6. A nationwide chain of retail clothing stores processes its daily sales transactions using a batch system based on a mainframe computer at a central location.

   a) Outline the flow of data through such a batch processing system. *(6)*

   b) The company is considering a change from a batch system to an interactive system. Describe the advantages and disadvantages of moving to an interactive system. *(4)*

   NEAB 1999 Paper 2

7. Graphic Designers are making increasing use of hardware and software systems to assist in the representation of pictorial and textual information on paper or on screen.

   a) State two benefits to a graphic designer of using these hardware and software tools. *(2)*

   b) While graphic design software can be used on a PC based system, additional or specialised hardware can be used to assist the designer. Specify four devices that might be used, stating clearly why they would be needed. *(4)*

   c) The completed image may need to be transferred to another package. One method is to store the image as a bitmap.

   d) i) What is meant by the term bitmap in this context? *(1)*
      ii) State three problems that could occur when a bitmapped image is used. *(3)*

   NEAB 1998 Paper 2

8. Graphical images can be stored in a bit mapped format. Explain what is meant by the term *bit mapped*. *(2)*

   NEAB 2000 Paper 2

9. State two types of data, other than alphanumeric, that can be stored in a computer file. *(2)*

   AQA ICT Module 2 May 2001

10. A college uses a computer-based batch processing system for keeping the students' records. The students provide their details, or changes to their existing records, on pre-printed forms. The completed forms are collected into batches ready to update the master files. These occur every night at certain times of the year, and once a week at other times.

a) Explain what the term *batch processing* means. *(3)*

b) i) Give **one** advantage of batch processing to the college. *(1)*
    ii) Give two disadvantages of batch processing to the college. *(2)*

The college decides to install a transaction processing system with which student records are keyed in on-line by a clerk.

c) Explain what the term transaction processing means. *(3)*

AQA ICT Module 2 Jan 2001

Dissemination and distribution

Information itself is very important, but the style in which it is presented is also important. Consider how two newspapers report the same story.

The style and content of computer output will depend on

- the target audience (which person or group of people is the information aimed at?)

- the purpose of the output.

Within an organisation there are likely to be a number of different kinds of audience for whom information might be required. The shareholders will wish to be given information summarising the performance of the organisation over the last year; a group of salesmen and women may need to be briefed about the prices and specifications of new products; an operations manager needs to be informed of the performance and output of each factory.

Output format

A sales manager is writing a report on her company's sales. She can use different formats to present the same information depending on who will read the report.

- A **director** may want to see sales figures at a glance, probably including last year's figures for comparison. A **column graph** like the one shown in Figure 17.1 could be suitable.

Pie charts (both 2-D and 3-D) are used to show the relative size of figures (see Figure 17.2).

Line graphs (Figure 17.3) are used to show trends in figures.

- An **accountant** may want to see more information, possibly broken down area-by-area, month-by-month and compared with previous figures. The information can be presented in a **table** and studied in depth. Part of it might look like Figure 17.4:

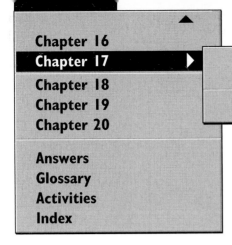

Figure 17.1 A column graph

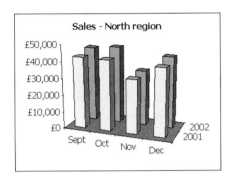

Figure 17.2 Pie Chart

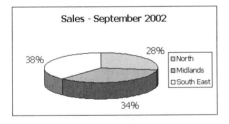

Figure 17.3 Line graph

2002	Sept	Oct	Nov	Dec
North	£45,126	£46,327	£33,112	£42,557
Midlands	£56,355	£48,181	£53,556	£52,489
South East	£62,318	£78,594	£69,298	£68,749
Total	**£163,799**	**£173,102**	**£155,966**	**£163,795**

2001	Sept	Oct	Nov	Dec
North	£42,433	£41,794	£32,036	£39,996
Midlands	£56,128	£50,982	£55,623	£56,634
South East	£61,760	£76,304	£66,050	£64,698
Total	**£160,321**	**£169,080**	**£153,709**	**£161,328**

Increase	Sept	Oct	Nov	Dec
North	6.3%	10.8%	3.4%	6.4%
Midlands	0.4%	-5.5%	-3.7%	-7.3%
South East	0.9%	3.0%	4.9%	6.3%
Total	**2.2%**	**2.4%**	**1.5%**	**1.5%**

Figure 17.4 A table

- The **manager's** immediate superior may want even more information. This may be an exception report including details of where the company has done exceptionally well or exceptionally badly. Reasons for good and bad performance may be included. A full text **report** may be necessary (Figure 17.5).

Sales figures in the Midlands region for the period October to December 2002 were particularly disappointing, each month showing a fall on the 2001 figures. During this time, one member of the Midlands sales team, Harry Brent, was on long-term sick leave. Harry has now made a full recovery and I expect the figures to improve during the next few months. In the rest of the

Figure 17.5 The full report

- A report to a **group of directors** could be presented on a screen using an LCD projector and presentation software such as Microsoft PowerPoint. Such a presentation could include graphs, tables and text. PowerPoint gives the option of printing out the presentation so that the audience have a hard copy of the report to take away with them for later reference.

WYSIWYG (What You See Is What You Get) software such as modern word-processors offer different formats, such as text, graphs and tables. Users can see how the output will appear before it is printed and make any necessary alterations (see Chapter 21).

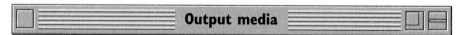

Output media

All the output formats described above can be produced on paper for distribution to the appropriate people. However, there are times when the information needs to be presented to a group of people at a meeting.

Figure 17.7 An LCD projector

Traditionally acetate slides with information printed (or written) onto them have been used with an overhead projector. These slides are cheap to produce and only require a standard overhead projector and screen for display.

Today many presentations are prepared using software such as Microsoft PowerPoint. Although individual slides can be printed out onto acetate and used as described above, the package allows for the information to be displayed as a slide show using an LCD projector and a screen.

An LCD projector is connected to a computer and the image normally appearing on the computer screen is projected.

Use of presentation software together with an LCD screen enables a range of extra features to be added. Animation can be used so that the contents of a slide can be built up bit-by-bit to tie in with what the presenter is saying. Sound and videoclips can be added to the presentation.

The speaker can use a remote control to move the slides on, and by using navigational tools provided by the software, can move onto different slides to suit the needs of the occasion.

The package will print out hard copies of the slides in a variety of formats so that members of the audience can have a copy on which to add notes. Such presentations appear much more professional than ones based on standard acetates.

A stand-alone slide show can be produced, stored on CD-ROM and distributed for use in a variety of locations.

> **Advantages of using presentation software** >

- 'Professional' looking presentation
- Speaker can be remote – slides can be changed using a remote control mouse
- Use of video, sound and animation improves impact
- Can produce printed copy of the slides in different formats
- Automation of presentation possible with pre-set timings between slides; different styles of transition from one slide to the next can be used
- Easier to keep in order – cannot drop slides

A multi-national company might prepare a slide show that presents details of new products. The show is to be used with sales staff. The completed presentation could be distributed on CD-ROM to different locations for use in simultaneous training sessions.

Electronic white boards

The use of electronic white boards together with LCD projectors is opening up further possibilities for presenters. As well as carrying out a presentation as described above, the use of an electronic white board allows the user to add notes to the image on the screen and store and print these annotations.

Formatting output

Design considerations for printed output

Once the type of output has been chosen, there are many techniques available to make information more readable.

- Text can be formatted to different colours and different fonts.
- Bullet points can make lists stand out.
- Bold print and formatting borders can make tables easier to read.
- In a formal report, sections and sub-sections can be numbered.

Design considerations for LCD projection

- Font size of text should be chosen so that it is large enough to be seen anywhere in the room when displayed on a screen.
- The amount of text shown on one slide should be kept to a minimum; key points should be given rather than lengthy sentences.
- Care should be taken over the use of colour. Some colours show up better than others.
- The background chosen should not be too cluttered otherwise it will distract from the text.
- Many organisations have standards that must be adhered to, for example the organisation's logo may need to be included in each slide
- The style, use of colour, the material and the level of language should be appropriate to the audience.

Dissemination on the Internet

The Internet has become a very important means for disseminating and distributing information. With so many web sites competing for our attention, many web pages (particularly those marketing a product) use different techniques to hold our interest, stop us moving to another site and persuade us to read the information.

Activity 1

Use a presentation package such as Microsoft PowerPoint to prepare a slide show to present a topic from the ICT AS specification to your class. Investigate as many features of the package as possible. Take care to ensure that your slide show keeps to the good design guidelines described above.

Activity 2

Choose any web site and make a list of the techniques used to hold the user's attention.

These techniques include:

- text colour and background colour
- marquee – text scrolling across the screen
- bright images
- sound
- animated gifs – apparently moving pictures
- animations using special software such as Macromedia Flash
- video

Style or substance?

When presenting information don't forget that the purpose is to communicate information. Whatever format is chosen, it is crucial that the design of layout is suitable for the given audience. Sometimes too elaborate a format will hide the content of the report. Stick to a clear design and avoid information overload.

Report generation

A report may be produced automatically using Report Generator software (see page 175). This software can take data from one or more files and paste it into a report in a pre-determined format. This format will include which fields to print and in which order. Calculations can be performed on numeric fields such as comparing sales for this month with last month or the same month last year. Reports can include graphs of numerical data. Once the report has been set up with one set of figures, the next report can be prepared automatically with the same format. Page formatting such as page numbering and orientation can also be carried out.

Summary

- the way in which information is presented will depend on the target audience
- different formats are available such as text, graphs and tables
- different formats and techniques can be used to make a report more readable
- information can be disseminated to a larger audience in one place by creating a slide show using presentation software and presenting it via an LCD projector
- different techniques are used on the Internet to attract and hold our attention

Dissemination and distribution questions

1. 'People remember 20% of what they hear and 30% of what they see, but 40% of what they hear and see.' Many business organisations are using presentation graphics packages to prepare and present material, often to a large audience.

a) Describe two major functional features of a presentation graphics package. *(4)*

b) Describe two different forms of output you would expect the package to produce to enable an effective presentation to be made. *(4)*

c) Describe one potential problem when displaying IT based information on a computer screen to a large audience, and explain how it can be overcome. *(2)*

NEAB Computing 1995

2. A manufacturing company intends to use an information system to store details of its products and sales. The information system must be capable of presenting the stored information in a variety of ways. Explain, using three distinct examples, why this capability is needed. *(6)*

1996 Paper 2

3. The head of a sales team has developed a presentation. It is planned for members of the sales team to deliver this presentation as part of a sales talk to large audiences at various locations throughout the country.

a) State three advantages to be gained by using presentation software as opposed to the use of traditional methods, e.g. OHP. *(3)*

b) State three design considerations that should be taken into account when the head of the sales team is developing the presentation. *(3)*

1999 Paper 2

4. A headteacher and the school's governing body want to consider the school's recent exam results at AS and A level at their next meeting. You have been asked for your advice on how the results should be presented. Produce a brief report on the options available. *(5)*

5. The head of a company's IT services department is to give a presentation on data security to all computer users within the company.

a) Give **three** methods of ensuring data security that she should include in the content of her presentation. *(3)*

b) She decides to develop a computer-based presentation to be displayed using an LCD projector rather than creating overhead projector transparencies.

i) State **three** functions of the presentation software that are only available for use with the LCD projector. *(3)*

ii) Describe **two** design considerations that she needs to take into account in order to develop an effective presentation. *(4)*

AQA ICT Module 2 May 2001

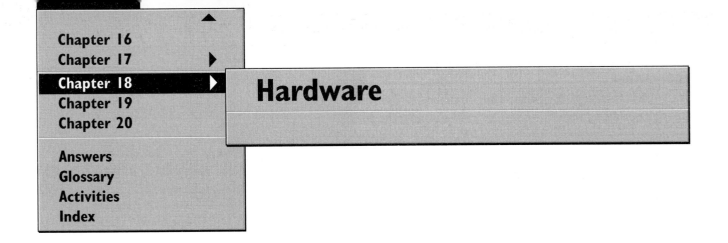

Hardware

Computers work on the basis of input, process and output. The various parts of the computer (the hardware) can be defined as input devices, the processor, backing store and output devices.

Input devices are used to enter data. The processor is in the 'box' part of the computer and is made up of electronic printed circuits and microchips. The processor includes the computer's memory. Backing store is where data is stored when the computer is turned off. Output devices present the information to the user.

Central Processing Unit

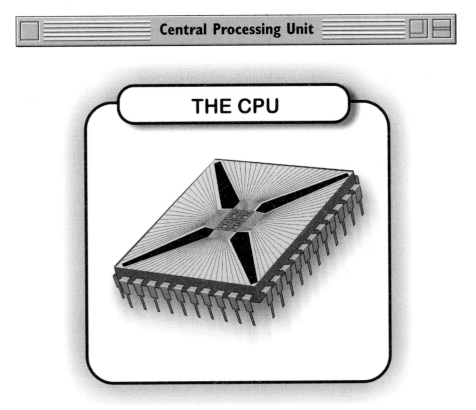

THE CPU

Figure 18.1 CPU

The Central Processing Unit (CPU) is where data is processed. When buying a computer you will see the clock speed of the CPU, measured in megahertz (MHz) or Gigahertz (GHz) advertised. Usually the faster the clock speed, the faster the data is processed. This is important when using some software that performs a lot of calculations, for example if it is graphics intensive.

The computer's memory is associated with the CPU. There are two types of memory chip: Read Only Memory (ROM) and Random Access Memory (RAM). Data in ROM cannot be changed and is permanently stored even when turned off. ROM is used to store the boot programs when the computer is switched on.

RAM is used to store any software and data while it is in use. The more sophisticated the software, the more memory it is likely to use. Multi-tasking – running several programs at once – demands a large memory. When the computer is switched off, data stored in RAM is lost.

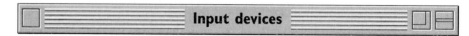

Input devices

(see Chapter 11 for more information)

The keyboard and mouse are not the only input devices. OMR readers, OCR readers and MICR readers are used by commercial businesses. Games computers use a joystick. Hand-held computers use keypads. Computers controlling manufacturing processes may use sensors.

The ergonomic keyboard is an alternative keyboard that is reputed to reduce the risk of repetitive strain injury (RSI). A concept keyboard is a simple keyboard for people with learning difficulties. Fast food chains use special keyboards where there is a key for each menu item.

Figure 18.2 A dedicated keyboard

Touch screens allow users to make selections by actually touching the screen. In fact there is a grid of infra-red beams in front of the screen. Pointing at the screen breaks the beams so giving the position of the finger. Touch screens are input and output devices combined. Palmtop computers use touchscreen and character recognition technology to allow the user to 'write' on the screen and the text entered into the computer.

Figure 18.3 A Scanner

Scanners can be used to scan in pictures or text. Pictures can be stored in a number of formats like jpg or bmp. Text can be stored as a text file and imported into other applications.

Hand-held scanners are cheap and suitable for home use, but the quality is not as good as flat-bed scanners (which have been reduced in price lately). Flat-bed scanners often come with OCR software, enabling text to be scanned in and stored so that it can be loaded into a word-processing program. The accuracy of this software depends on the quality of the original text.

Digital cameras have reduced in price considerably and now cost little more than conventional cameras. Some digital cameras store pictures in jpg format directly onto a conventional floppy. Other digital cameras connect to a computer's serial port so that pictures can be downloaded. It is possible to imagine that conventional film will become obsolete in the near future.

Graphics tablets have a stylus pen which is used to draw on a special flat surface. The drawing is automatically read by the computer. These are ideal for graphic designers, artists and technical illustrators.

Output devices

Output devices include the monitor, printers, plotters and loudspeakers.

The most common forms of printers all form their images out of dots. The smaller the dots, the better the quality of the print.

- **Dot matrix printers** print by hammering pins against a ribbon onto the paper to print the dots, so they can be noisy. They are cheap and although quality of print and noise levels have improved, dot matrix printers have lost popularity as the price of ink-jet and laser printers has fallen.

- **Ink-jet printers** squirt ink on to the paper and form letters from tiny dots. There are quiet, quality is good and colour versions are available for as little as £100.

- **Laser printers** are the fastest and produce the best quality print but are also the most expensive type of printer. Prices have dropped and they are now on sale for little more than £250.

- **LCD projectors** which project computer output onto a large screen are expensive but are now common in business demonstrations projecting the output of a laptop computer. Prices continue to fall and quality is improving.

- The use of a loudspeaker that allows for speech synthesis is an expanding area of output, common in computer games. If you phone up directory enquiries or the speaking clock, you will be told the number you require or the time by the computer in synthesised speech. It sounds almost human.

Backing store

Backing store is a permanent storage medium on which data can be stored for retrieval later. Usually backing store is a **magnetic disk** although in certain circumstances **magnetic tape** is used.

Tape is much slower as it only offers serial access – it has to be read in the order in which it is stored on the tape. Access times for data held at the wrong end of the tape are very slow. Today tape is mainly used for back-up where the whole disk is copied onto tape.

Personal computers use two sorts of magnetic disk.

● The **hard disk** inside the computer can hold many gigabytes. It is used to store the operating system, software and datafiles. These files are vital and some form of back-up is necessary in case of disk failure. Although it is possible to have a removable hard disk for better security, these are only used rarely. Normally the hard disk is fixed into the computer.

● The removable 3.5' **floppy disk** normally stores only 1.44 Mb and is much slower than a hard disk. It is mainly used for keeping back-up copies of small files or for transferring small files between two or more computers. Floppy disks are also used to hold control information, such as printer driver 'fixes'. The procedure that allows a computer to be 'booted up' when a new operating system has to be installed is usually held on a floppy disk. Floppy disks are very cheap.

'SuperFloppy' disk drives can store up to 120Mb on one special floppy disk. This is the capacity of over 80 floppy disks. They are five times faster than the standard floppy drive. They can be used for back-up, but several SuperFloppies would be needed to back-up the contents of a hard drive.

CD-ROM

CD-ROMs (Compact Disk-Read Only Memory) are small plastic optical disks. Like floppy disks, CD-ROMs can be moved from computer to computer. However, they store large quantities of data (650 Mb or 0.65 Gb) permanently with short access times. A CD-ROM drive of some sort is now standard in PCs. The data stored on a CD-ROM is less prone to damage than that on a floppy disk as the data storage method uses laser rather than magnetic technology.

Most software is distributed on CD-ROM and many computing magazines are sold with a CD-ROM attached that holds free sample software.

Figure 18.4 CD-ROM

CD-ROMs are used to store multimedia applications as the sound, graphics and animation files are likely to require large amounts of storage space.

CD-Rs (Compact Disk – Recordable) are writable compact disks. Using a special writable CD drive, up to 650 Mb of data can be recorded on a blank disk. This data cannot be altered but it can then be read by any other PC with a standard CD-ROM drive. This means that CD-R is suitable for back-up as it holds much more than a floppy disk, yet can be taken off site easily.

CD-RWs (Compact Disk – ReWritable) are re-writable compact disks. Using the same writable CD drive, up to 650 Mb of data can be recorded, deleted and re-recorded on these disks. This means that a CD-RW can be used in the same way as a floppy disk, but obviously storing far more data.

DVD

DVD (Digital Versatile Disk) is a high-capacity optical disk developed in the 1990s. A DVD is the same size as a CD. It is expected that DVD will replace CD-ROM and videotapes for films.

DVD drives can read both CD-ROMs and DVDs. 'Combi' drives that read DVDs but can write to CD-R and CD-RW are becoming increasingly common. Some DVD drives can both read and write to DVDs.

The highest capacity DVD version can store up to 17 Gigabytes – the equivalent of over 25 CD-ROMs. DVD-R (Recordable DVD)and DVD-RW (Re-writable DVD) are available but capacity is reduced.

Magnetic tape

The traditional large, reel-to-reel magnetic tape is rarely seen today, However, a variety of types of tape cartridge are now used for backing-up files. These are discussed in detail in Chapter 19.

CD-ROM or floppy disk?

CD-ROMs and floppy disks can both be easily removed from one machine and used in another. They are therefore both suitable for transferring files from one machine to another or for installing software.

What are the main differences between floppy disks and CD-ROMs? When should a floppy disk be used and when should a CD-ROM be used?

The main differences are as follows:

- CD-ROMs have a much higher capacity.
- CD-ROMs are much faster than floppy disks.
- A floppy disk can be used over and over again as files can be deleted and new files added. A CD-ROM is read-only, however CD-RW can be re-written to and CD-R can be written to once.
- CD-ROM drives are more expensive to buy than floppy drives but CD-ROMs are cheaper per megabyte than floppy disks.
- Floppy disks are magnetic media, while CD-ROMs use a laser scanner.
- Floppy disks are easily damaged, while CD-ROMs are not.
- Floppy discs are suitable for storing small datafiles, for example for a student to transfer work from college to home.

CD-ROMs are widely used to distribute software as the installation files will all be stored on a CD-ROM. This is because of the large capacity and because they are less likely to be damaged. Multimedia applications files storing sound, video, animation and text, for example for an encyclopaedia, are likely to be comparatively large. They will probably need to be stored on CD-ROM.

Miniaturisation

As microelectronics has developed we find that electronic hardware is getting smaller. Mobile phones, MP3 players, digital cameras, video cameras and even alarm clocks have reduced in size. These are all examples of **miniaturisation**. Pocket computers (palmtops) can load spreadsheets and word-processing software and access the Internet despite being little bigger than a small diary.

Figure 18.5 Palmtop

Summary

- computers consist of input devices, the processor, output devices and backing store
- many different input devices are available in addition to the most common mouse and keyboard
- miniaturisation is developing all the time
- output devices are not just the monitor and the printer
- speech recognition and speech synthesis will continue to expand, as they get more reliable

Hardware questions

1. Write a brief paragraph on each of the following stating when it is likely to be used and why.

    a) Ergonomic keyboard

    b) Concept keyboard

    c) Touchscreen

    d) Voice recognition

    e) Graphics tablets

    f) Laser printers

    g) Dotmatrix printers

    h) Ink-jet printers

    i) Speech synthesis

2. Why are loud speakers now common as computer output? *(2)*

3. A company sells a range of health foods at five different shops. It also sells directly to the home from a number of vehicles. There are hundreds of different items of stock and many items are seasonal, so items in stock are constantly changing. Customers purchase goods and pay by cash, cheque or credit card.

The company is considering a computerised system to help manage sales and stock control. Bearing in mind the needs of this company what are the capabilities and limitations of possible

    a) communications devices *(4)*

    b) input devices *(4)*

    c) output devices *(4)*

    d) storage devices *(4)*

        based on a question from NEAB 1997 Paper 2

4. Modern personal computer systems usually include a floppy disk and a CD-ROM drive.

    a) List three main differences between these two devices. *(3)*

    b) i) Describe one example of an application for which floppy disks are used. *(2)*

        ii) Describe one example of an application where a CD-ROM is used. *(2)*

        NEAB 2000 Paper 2

5. State a suitable medium for transferring sound files, and give a reason for your choice. *(2)*

    AQA ICT Module 2 January 2001

6. A new printer is supplied with printer driver files. The files are provided both on a floppy disk and on a CD-ROM. The CD-ROM also contains sound files for use with the printer.

    a) State one reason why the sound files are not provided on a floppy disk. *(1)*

    b) Give one possible use of the sound files. *(1)*

    AQA ICT Module 2 May 2001

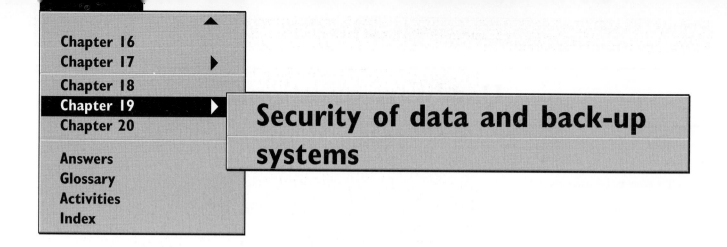

Security of data and back-up systems

Data stored on computer is vital to the success of any business. The loss of computer files is an extremely serious problem for any organisation, so it is vital that businesses take steps to protect the security and integrity of their data.

- **Data integrity** means the correctness of the data both during processing and when it is stored. Data may be incorrect because of errors in data transmission (caused by background noise on the line), input errors (data typed in wrongly), operator errors (for example an out-of-date version of the file has been loaded), program bugs, hardware breakdown, viruses or other computer crime.

- **Data privacy** means keeping data secret so it cannot be accessed by unauthorised users. The Data Protection Act (see Chapter 9) provides legal protection for the data relating to an individual. It is necessary to establish system checks to maintain data privacy and reduce the risk of unauthorised access.

- **Data security** means keeping data safe from physical loss as well as confidential and correct. Loss could be due to accidental damage, for example natural hazards such as flooding and fire, or it could be damage caused by hardware failure, for example when a tape gets caught up in a drive and is destroyed. The loss of data could be intentional, for example theft by a competitor, unauthorised access (hacking), destruction by viruses or even terrorism.

If a computer user orders goods by e-mail, they need to give their credit card number and expiry date. This is **private** information, which someone else might use to order goods fraudulently if the information was not kept secret.

Activity I

213

For each individual citizen, large quantities of data are stored in the computer systems of many different organisations.

○ Copy out the table below. For each entry list the types of data that would be stored.

Example	Type of data
Credit card transactions	Card number, amount of purchase, shop name, date, time
Electoral roll	
Utilities (electricity, water, gas, telephone)	
Mortgage and loans	
Banks	
Employer	
Doctor, dentist, hospital	
School, college	
Clubs and societies	
Store accounts	
Loyalty cards	
Mobile phone	
Driving Licence	
Vehicle Registration	
Insurance companies: life, household, health, car, holiday	
Bank cash dispensers	
Internet shopping	
Market research questionnaires	

○ For each category discuss the ways in which the data could be misused.

Data integrity is lost when data is altered in some way, making it incorrect. Such alteration can occur accidentally, or through malicious intent. A number of measures to protect systems from illegal access such as data encryption and virus checking that were discussed in Chapter 7 are appropriate here. A number of further measures can be taken to help minimise the risks to data integrity.

Standard clerical procedures

Loss of data integrity often occurs not through computer malfunction, nor as a result of illegal or malicious access, but as a result of user mistakes. For example, yesterday's transaction file could be used to update a master file instead of the correct one for today. This could result in yesterday's transactions being carried twice and today's not at all. To ensure that such errors do not occur, very careful operational instructions should be laid out and enforced. Files should be properly labelled and stored in a systematic, pre-determined and clear manner. Detailed manual records of the location of files should be maintained.

Write protect mechanisms

Data can mistakenly be overwritten if the wrong disk or tape is used. Floppy disks are designed with a special slider mechanism. When the slider is moved to a particular position writing to the disk is prevented. Similarly certain tapes require a plastic ring to be inserted before the tape can be written to. Care should be taken to write protect any disk or tape that is storing data that needs to be preserved.

Passwords

In a networked system only registered users will be allowed access. Each authorised user will be allocated an individual user identification code that they will enter to log on to the system. They will be asked to enter a password to confirm that they are indeed the identified user.

Passwords need to be kept private otherwise they have no value. They should be carefully guarded and never revealed to others. A user should take care over her choice of password. A password should not be easy to guess. For example, names should be avoided, as should words such as SECRET, SESAME, KEEPOUT. A password should not be too short otherwise it can be easily decoded. Ideally, it should not be a real word but simply a collection of characters, perhaps a mixture of numbers and letters. Some passwords are case sensitive: if so it is a good idea to mix up upper and lower case letters.

It is essential that a password is never written down; far too often users write their password down in their diary, or on a piece of paper which is kept in an easily accessible desk drawer. Even worse is to write the password on a sticky label that is stuck on the screen of the computer!

Levels of permitted access (see Chapter 7)

Not all users need to be able to access all files on a networked system. Access rights can be set up that only allow certain users to have access to specific files or applications. Not all users need to access data in the same way as there are a number of different levels of access that can be permitted. These include:

- **Read only**: the user can view the data in a file but not alter or delete it

- **Read/write** access: the user can modify data as well as view it

- **Delete** access: the user has the authority to remove a file or record

- **No Access**: a user cannot access the file in any way.

Back-up and recovery

The value of data to an organisation can far exceed the value of the physical computer system. Loss of this data could lead to the collapse of the business. A study by the University of Texas Centre for Research on Information Systems showed that businesses never recover from a loss of computers that lasts for ten days or more. The same study found that 90 per cent of data losses were due to accidents such as power failures, water leaks, loose cables, user mistakes, and other hardware, software and human errors.

When the term backing-up is used, it generally refers to the process of copying files. The purpose of back-up is to ensure that if anything happens to the original file, the back-up copy can be used to restore the file without loss of data and within a reasonable timescale. Thus back-up is used to avoid permanent data loss and ensure the integrity of the data.

It is often hard to imagine the scale of the problems that can occur when data is lost. A company that loses all its data relating to sales, customers and suppliers for example, could quite possibly be unable to continue trading and hence go out of business. As the Texas study found, the timescale in which lost data can be restored can also be a crucial factor in an organisation's survival.

The backing-up of files is not enough by itself to protect a computer system from huge data loss. An organisation needs to put procedures in place that will allow the lost or corrupted files to be restored by making use of the back-up copies. These procedures need to be carefully planned and personnel made aware of them; the methods of recovery should be practiced so that, when they are needed, they will run smoothly.

While emphasis is placed on the need to back-up data files, there are other electronically stored files that are crucial to the running of a system.

Without an operating system a computer will be virtually unusable. A modern operating system is complex and requires many stored files. Many of these are specific to the installation and include appropriate device drivers, fonts and control panel settings. Without suitable back-up, it would be difficult and very time-consuming to restore the environment to its original state if the files were corrupted.

Applications, though usually installed from CD-ROMs these days, are usually customised once installed to meet the specific needs of the user. Without back-up, all such customisation could be lost.

In batch processing (see Chapter 16), when a master file is updated the old version is still intact at the end of the process. This provides an automatic back-up file. In all other processing modes data is overwritten as transactions occur, so back-up copies of the file will need to be made.

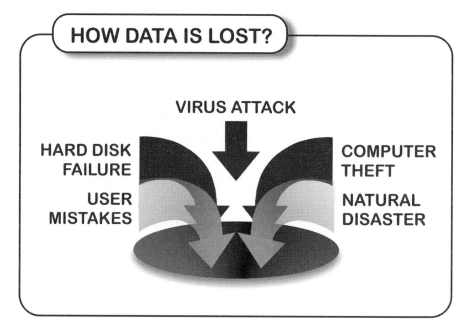

HOW DATA IS LOST?

VIRUS ATTACK

HARD DISK FAILURE

COMPUTER THEFT

USER MISTAKES

NATURAL DISASTER

Figure 19.1 How data is lost

For an individual user, working on a PC, there are a number of measures that can be taken to make sure that, in the event of data loss, important files can be restored.

Many software packages have automatic back-up procedures. Typically they could be **back-up on saving** where the old version of a file is saved whenever a new version is created. The old version is automatically renamed before the new one is saved. The old file is usually given the same name but an extension BAK. For example, a file named Letter.DOC would be backed up as Letter.BAK while the new, updated version of the file would replace the old Letter.DOC.

A second method of automatic back-up is the timed back-up that is useful in case a power failure occurs. The user sets a time period, such as five minutes, after which the document is saved automatically. Many users do not make use of these facilities that are offered by software packages. They may find the time taken to back-up lengthy documents irksome or feel that they do not have adequate storage space to keep extra copies of files.

Careful use of such automatic back-up facilities will not protect a user from complete hard disk failure, an occurrence that is not unusual. To protect against such a failure, a user should copy files to a different storage medium such as floppy disk.

For larger, networked systems the back-up issues are more complex. Some data will be stored centrally, on network file servers, while other data may be stored by an individual user on the hard drive of their own workstation. Very often the user will be expected to manage the back-up of their own, locally held files, while the network manager will take responsibility for all centrally held data. A number of factors need to be considered when determining an appropriate back-up regime. These factors are discussed below.

Selection of appropriate medium for back-up storage

When choosing the appropriate medium for back-up the capacity (the amount of data that can be stored on the medium), the speed of transfer of data and the cost will all need to be considered. The hardware device used can be dedicated solely for back-up use or used for other purposes as well.

Some examples of dedicated back-up devices

8mm tape	Offers moderate speed (60 –80 MB per minute) and good capacities (2.5 to 40GB) for medium sized environments.
ADR (Advanced Digital Recording)	Offers high speed (30 – 120 MB per minute) and good capacity (50 GB) for desk top computers and servers.
AIT (Advanced Intelligent Tape)	Capacity up to 50 GB and transfer speeds of up to 360MB per minute make AIT suitable for environments with large amounts of data.
DAT (Digital Audio Tape)	Holds 12 – 20 GB with transfer speeds from 6 to 150 MB per minute.
QIC (Quarter Inch Cartridge)	QIC drives suitable for home or small business use being relatively inexpensive. Cartridges can be bought with capacities from 4 to 50GB and speeds can range from 30 to 300MB per minute.

A tape library is a device that makes unattended large-scale back-ups possible. They are frequently used when backing-up data stored over a network. A tape library consists of tape drive and a robotic device that can move tapes in and out of the drive. By using a tape library, the volume of data that can be backed-up without the need for human intervention is increased hugely and limited only by the number of tapes that it can hold.

It is not always necessary to have a device that is dedicated to back-up. Other storage devices that may also be used for archiving or transferring data may also be suitable for back-up.

Floppy disk	Very low capacity; suitable for storing a few files.
CD-R and CD-RW (Compact Disk Recordable and Re-writable)	Store 1.2 GB on each disk with transfer speeds from 5 too 50 MB per minute.
DVD-RAM (Writeable Digital Versatile Disk)	Capacity up to 9.4 GB on a double side cartridge with speeds up to 60MB per minute.
Removable Cartridge Devices	SuperDisk, Zip and Jaz drives are good for personal back-up use as well as for transferring files between different computers. Capacities range from 100MB to 2.2GB with transfer speeds from, 30 to 480 MB per minute.

Frequency of back-up

The frequency of back-up must be determined. Obviously, the more frequent the back-up, the less out of date will be the data when it is restored. However, whenever a back-up is undertaken, processor time is tied up and files can be unavailable for other use. An appropriate balance needs to be found and factors such as the acceptable length of delay in restoring files in the case of failure as well as the importance and nature of the data. Sales data for a supermarket, which affects orders and deliveries, will be backed-up hourly, if not more frequently. User data, such as passwords and user names, need only be backed-up every week.

Recording of transactions in a log

Unless a file is backed-up after every transaction, which is most unlikely to be feasible, a record will need to be kept of all the transactions that have taken place since the last back-up occurred. In the case of file failure, the latest back-up copy would be used to restore the file. It could then be brought up-to-date by re-running all the transactions stored on the log, that had occurred since the back-up was made.

Full or differential back-up

A back-up takes time. If the contents of a hard disk are backed-up every day, many of the files that are backed up, probably the majority, will not have been altered since the last back-up was taken. To keep re-backing-up the same data again after day wastes both time and back-up media space. A back-up where all files are copied is known as a **full** (or **global) back-up**. An **incremental back-up** only copies files that have been created or changed since the last back-up was made. A common back-up regime might involve an organisation making a full back-up once a week with incremental back-ups made nightly.

Back-up copy generations

As back-up copies of files can themselves be damaged it is necessary to keep a number of old back-ups. These are known as generations.

Recovery method

It is important that when files have been backed-up that the back-up copy is verified, in other words checked against the original to ensure that it has been copied exactly. If this is not done, the back-up files could prove to be useless and the original file could not be recreated.

If the original data files are lost or corrupted, the data can be recovered by using programs that restore the data from the back-up files. It is necessary to restore the files in the correct order by following agreed recovery procedures. The file will first be recreated using the most recent full back-up and then each subsequent incremental back-up file should be accessed, in time order, to update the file. Any transaction log should then be used to restore the most recent transactions.

If the files are used in the wrong order, the restored file will not be correct. Care must be taken in the careful labelling and organisation of back-up tapes and disks to ensure that no mistakes are made.

Physical security of back-up medium

As data can be lost due to disasters such as fire, it is necessary that back-up files should be kept in a separate place from the original file. Often a fireproof safe is used.

Case Study 1

A Remote Back-up Service

Mercer DataSafe is a company that offers businesses an automated off-site back-up service. Data from the company's computer is automatically backed-up after business hours to Mercer DataSafe's computer in Lawrenceville, New Jersey, USA.

Using Mercer DataSafe's back-ups, businesses can quickly recover any lost data, usually within 24 hours of a catastrophic loss (such as a flood, fire, theft or operator error), or within hours of a minor loss (such as a deleted file).

To use Mercer DataSafe's system you only need a modem. At the pre-set time, the data to be backed-up is compressed in a securely encrypted format. Mercer DataSafe's computer in Lawrenceville is contacted using the modem. After identification checks, the data is transmitted to the Lawrenceville office.

When finished, the data is verified. If everything is correct, a confirmation message is sent back to the company's computer and the computer shut down. The back-up company won't have access to the user company's records as all files are securely encrypted and only the user has the password.

Users who already have a tape back-up system may still find it better to use a remote service. Many users forget to perform a back-up even though they have a tape drive. Other users think they may be doing a back-up correctly, but when they need to restore a file they find out that their tapes are useless. Few users take their tapes off-site so, if they have a fire or other disaster, they lose all their data.

- What is meant by data verification?
- Why is the data compressed?
- Why is the data encrypted?
- List the advantages to an organisation of using the services of DataSafe.

Activity 2

All too often an ICT student who is approaching a project deadline reports that all his work has been lost due to hard disk failure. Suggest in detail the steps that a student (and you) should take to ensure that work is not lost in this way.

Summary

Data is a most valuable commodity to organisations.

Data can be lost because of

- disk failure
- user mistakes
- computer theft
- virus attacks
- natural disasters

Backing-up refers to the process of copying files.

Some software packages offer automatic back-up facilities such as

- back-up-on saving
- timed back-up

The following factors need to be considered when establishing a back-up regime:

- the appropriate medium for back-up storage
- the frequency of back-up
- the use of a log to record transactions
- the use of full (global) or and incremental back-ups
- the number of generations of back-up that should be kept
- the method of recovery
- how the physical security of back-up medium is to be assured

Security questions

1. A computer system that is normally in use 24 hours a day holds large volumes of different types of data on disk packs. The main types of data stored are:

- applications software that changes only occasionally during maintenance
- data master files that are updated regularly every week
- transaction files which are created daily
- database files which are changing constantly.

It is vital that these different types of files can be quickly recovered in the event of file corruption. Outline a suitable back-up strategy for each of these types of files explaining what data is backed-up and when, the procedures to be followed and the media and hardware needed. *(8)*

NEAB Specimen Paper 2

2. An Internet sales company carries out its business with the assistance of a database system running on a network of PCs. The main tasks are the processing of customer orders and the logging of payments. You have been asked to advise the company on back-up strategies and to explain their importance.

a) Give two reasons why it is essential that this company has a back-up strategy. *(2)*

b) State five factors that should be considered in a back-up strategy, illustrating each factor with an example. *(10)*

AQA ICT Module 2 Jan 2001

3. Employees can often be responsible for causing loss or damage to their company's data. Regular back-ups are taken by the company, but in order to prevent employees from causing such loss or damage, describe:

a) two measures that could be incorporated into the hardware used; *(4)*

b) two software features that could be used; *(4)*

c) two other procedures that the company could introduce. *(4)*

AQA ICT Module 2 Jan 2001

4. A publishing company administers its business by using a database system running on a network of PCs. The main uses are to process customer orders and to log payments. You have been asked about back-up strategies and their importance.

a) Give two reasons why it is essential that this company has a workable back-up strategy. *(2)*

b) State five factors that should be considered in a back-up strategy, illustrating each factor with an example. *(10)*

c) Despite all the precautions, some data might still be lost if there was a system failure. Give two reasons why this might be the case. *(2)*

AQA Specimen Paper 2

5. The owner of a small newsagents uses a computer to manage her orders and deliveries. Every week she copies the files onto a number of floppy disks and puts the disks into a drawer next to the computer.

a) State **three** problems that may be caused by this method of back-up. *(3)*

b) Describe a more appropriate back-up procedure. *(6)*

NEAB 2000 Paper 2

6. The head of a company's IT services department is to give a presentation on data security to all computer users within the company.

a) Give three methods of ensuring data security that she should include in the content of her presentation. *(3)*

AQA ICT Module 2 June 2001

Network environments

What is a computer network?

A computer **network** consists of two or more access points called network stations that are linked together. Network stations are usually computers but might be 'dumb' terminals which have no processing power of their own.

A computer that is not part of a network is called a **stand-alone** machine. Stand-alone computers can only access datafiles stored on that computer. They will need their own printer and other peripherals such as a scanner. Datafiles stored on networked computers can be accessed by different network users. Networked computers can share **hardware** (such as printers or CD-ROM drives) and **software**.

Advantages of networks

- Hardware resources such as printers can be shared, so saving money.

- Datafiles can be shared rather than every user having their own copy of the file. This means that if the file is updated, everyone has access to the latest information.

- Communication between network stations is possible. This is particularly useful for e-mail and sending data accurately so that it does not have to be typed into the computer again (for example, examination board entries, newspaper stories and National Lottery tickets sales).

- Access to software and data files can be controlled. Different users can have different access privileges.

- You will only need to install the software once and it will be available to all network stations. With 20 stand-alone machines the software will need to be installed twenty times.

- When buying a network of say 20 machines, you can buy a network licence to run a program such as Microsoft Office. This will be cheaper than buying 20 copies of Microsoft Office, which you would need to do if you had 20 stand-alone machines.

- Security and back-up can be controlled centrally.

> ## Disadvantages of networks

- If the network server fails then it is likely that every network station will not work.

- Security may not be as tight if a network has remote access to a wide area network such as the Internet.

- A network may be slow depending on the amount of network traffic and speed of connection.

- A virus introduced on one network station may quickly spread to the rest of the network.

- Cabling may be difficult and expensive to install.

LANs and WANs

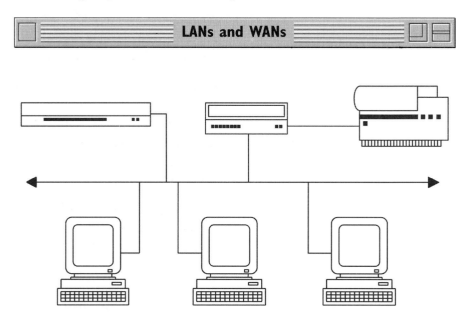

Figure 20.1 Computers linked in a LAN

A network may be in one room or one building or cover a small geographical area. This is called a **Local Area Network** or **LAN**. It will normally use its own dedicated cables.

Alternatively a network may be spread over a wide geographical area, possibly covering different countries. This is called a **Wide Area Network** or **WAN**. It can be linked by dedicated cables or by public telecommunications systems such as telephone lines.

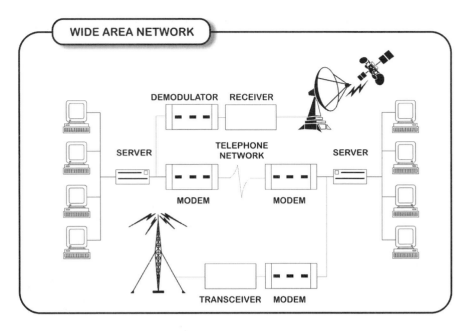

Figure 20.2 Computers linked together over a wide area

Networks provide opportunities for communication, particularly the sharing of datafiles and e-mail. However the greater the number of access points, the greater the problems in terms of security. These problems might be caused by

○ remote users – a WAN system that can be accessed away from the site will not be as secure as one than can only be accessed on-site such as a LAN system

○ unsupervised users – users accessing a WAN system from home will not be supervised in the same way as if they were in an office with other employees

○ unauthorised access – if a WAN system can be accessed legitimately from a remote computer, there is a danger that an unauthorised user may be able to gain access

○ difficulty in tracking down abuse – it is not easy to trace where unauthorised access has been made

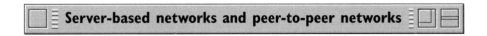

Server-based networks and peer-to-peer networks

There are two different types of local area networks.

1. **Server-based LANs** have a central computer called a server. Data files and software are usually stored on the server but can be accessed from network stations. As software is installed centrally, it only has to be installed once. All files are stored centrally providing a pool of data that is accessible to all workstations on the network. Back-up is easy to perform and there is no need to rely on users backing up their own files.

Dedicated servers provide user workstations with fast access to network resources (software, data and peripherals).

Individual users are set up centrally with appropriate access rights. Each user is allocated a user name, a password and disk space on the server. Security is therefore very high.

This sort of network is heavily dependent on the server. Servers need to have fast processing speeds, large memory and large hard disks. They are expensive and server-based networks are complicated to install.

Large LANs may have more than one server. The network can support computers of differing type.

2. **Peer-to-peer LANs** have no central server as all stations on the network are equal. Installing software takes more time, as it has to be installed on each station. As there is no central server, datafiles can only be stored on the network computer's hard disk. Backing-up must also be done separately for each individual station.

   Stations on a peer-to-peer network can access worked stored on that station and depending on privileges can access data stored on other machines. Such a network cannot have complete security. It would not be suitable for use in a school for pupils to use!

   As a server is very expensive to buy, a small peer-to-peer network would probably be much cheaper than a small server-based network. A peer-to-peer network would be ideal in a small office where four PCs need to be networked to share data.

Both types of LAN can share printers and other peripherals such as scanners. Both types of network can be used to send and receive e-mails from other network users.

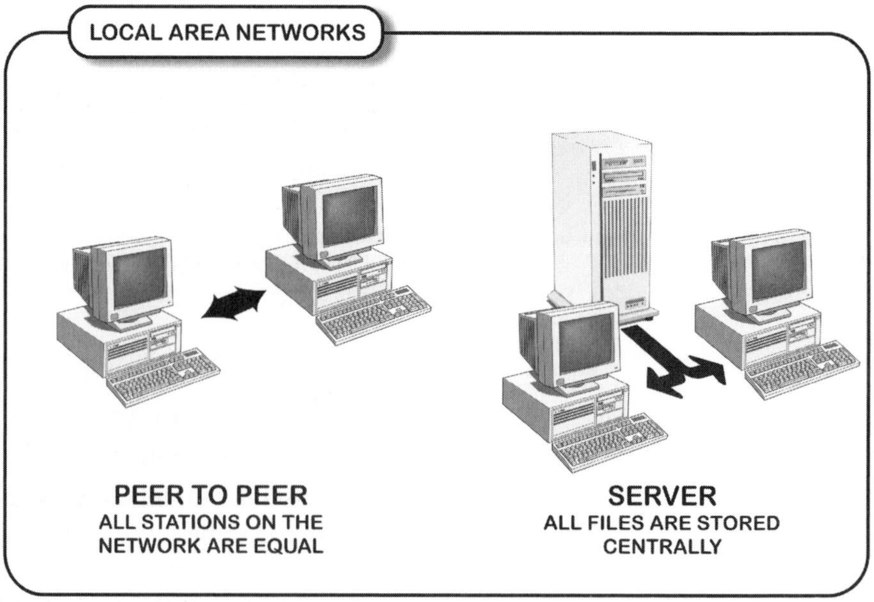

LOCAL AREA NETWORKS

PEER TO PEER
ALL STATIONS ON THE
NETWORK ARE EQUAL

SERVER
ALL FILES ARE STORED
CENTRALLY

Activity I

Find out what sort of network your school/college uses. How many servers are there?

Figure 20.3 Differences between two methods of linking a network

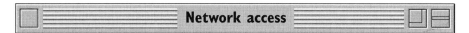

Network access

It is important that a network offers different users **different levels of access**.

The network manager will have unlimited access to all areas and drives. The manager will need greater privileges than ordinary users, for example in order to install new software, add and delete users, set up menus and so on. The manager's programs obviously must be protected by passwords.

Users, however, need read/write access to a dedicated area of the disk where their files are saved, read-only access to some areas (for example where software is loaded from) and no access at all to other areas (for example other user areas).

Case Study

Levels of Access

Allan Bromley is the network manager for a small plastics manufacturing company. The company's financial records and payroll details are stored on this network. Unauthorised personnel are not allowed to access this data.

○ Allan and other authorised users can inspect the data.

○ Authorised users add to the data, for example when someone starts work at the company.

○ Authorised users can edit the data, for example when employees get a pay rise.

○ Authorised users can delete some of the data, for example when an employee leaves.

○ Allan backs up all the data every night on to magnetic tape.

○ Allan safeguards the data by locking the previous day's back-up tape in a fireproof safe.

The growth of Wide Area Networks

There has been a great increase in the use of WANs over the past ten years due to

○ the increased importance of the Internet for information and e-commerce

○ the increased importance of e-mail as more people have e-mail addresses

○ the need to share information

○ the increase in information available

○ the increased speed of transmission and access

○ the fall in charges to access the Internet

○ the ability of computers to store more information.

Examples of the use of WANs include airline, theatre and hotel booking systems; home banking; bank ATMs; the Internet; the National Lottery; videoconferencing; stock control; e-mail.

Elements of networks

Important parts of networks

Network card	If you want to connect a computer to a local area network you will need a **network adaptor**, sometimes called a network interface card or **network card**. The card is an electronic printed circuit that fits inside the computer into an empty slot on the computer's motherboard (main printed circuit). The network cabling connects to the network card.
Cabling	Cables are the most common form of transmission media used to connect network stations to the rest of the network. They are usually made of copper wire. Common examples are **Ethernet** cabling (co-axial like a television aerial) and **UTP** (Unshielded Twisted Pair) cabling. Fibre optic cables are becoming more common; they are faster than copper cabling but more expensive.
Repeater	Transmitted signals deteriorate as they travel distances until they reach a stage when they are unrecognisable. To prevent this happening, a repeater is installed between segments in the network. A repeater is an electronic device that receives a signal and outputs it in boosted form.
Hub	The cable from the network station will connect to a hub. A hub is a switching cabinet that will connect to the server and to several stations. A local area network may have several **hubs** depending on the geographical location and number of stations. A hub can act as a repeater and boost a signal. If one station crashes the hub will ensure that the other stations work normally.
Bridge	A **bridge** is a device, which allows two local area networks to be linked, thus extending the network. This may be necessary if a network has reached a maximum size, either in the number of terminals covered or in the distance covered. An intelligent bridge can enhance security by only allowing messages from designated workstations to cross the bridge.
Backbone	If a number of networks are to be linked, a **backbone** can be used. This is a length of cable to which the networks are linked, each via a bridge. Each segment would have its own server. Users would have access to other networks and their servers via the backbone.
Network gateway	A **network gateway** enables a LAN to connect to a wide area network such as the Internet. This enables local area network users to be able to send e-mails to and receive them from people outside their LAN and to access Internet pages.

Wireless networks

Cables are no longer necessary to connect network stations. A wireless network does not need cables. At present wireless networks are not as fast as conventional networks but are very useful when

- connecting portable laptop computers to a network

- operating in temporary buildings

- operating where conventional cabling is impossible to install

Wireless networks require an **antenna** or transmitter, which transmits to a **wireless network card** fitted into the computer. Antennae have a limited range and a line of sight may be required.

Case Study 2

Computing 6 July 2001

Working on a Wireless World

While offering higher flexibility, wireless networks have been held back by a lack of speed, standards and reliability. Only now is the wireless local area network (LAN) beginning to enjoy industry respect after its jump in data rate from 2 Mbps to 11 Mbps and the Wi-Fi standard for compatibility.

Broadband fixed wireless is a suitable alternative offering high enough data rates to compete with, and improve upon, the technologies that we rely on today.

Wireless has the additional advantage that cables do not need to be laid, making the technology cheaper and easier to install with little maintenance.

Fat is the way forward

Until recently, there was no real need for fat pipes to every building, but with the growing amount of content online and the ever-increasing impact of e-business, this has changed. Although fibre to the building would be preferable in most cases, where this cannot be done – for any number of reasons – a wireless link could be the way forward for a company.

This could be, for example, where installation of cabling would inconvenience everyone, by digging up a busy street, for example. In cases such as this, a microwave transmitter can be used to provide a high-speed link at a fraction of the cost.

The only problem with this technology is that it relies on line of sight to work properly. Without this, a connection may still be made, but over time it will be unreliable and the signal will become unstable. This will typically give ranges of five to 30 miles depending on the local topography.

A saving grace of the technology is the price. A licence has to be bought to operate a fixed wireless connection, but the rental is typically a few hundred pounds. When compared with the cost of a fixed line, the wireless option presents quite a saving, provided that the line of sight requirement can be met. The technology effectively bypasses the local operator in a bid to keep down the cost of networking.

Speed is still often a limiting factor in wireless networks, and while the industry is rushing to play catch-up, wireless is still a long way behind the cabled world.

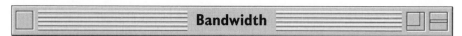
Bandwidth

The speed at which a networked computer can receive data depends on the bandwidth. A home computer connected to the Internet through a modem can only operate at the speed of the modem typically 56 Kbps (Kilobits per second). In fact, error checking procedures mean that even these relatively slow speeds are not achieved. ISDN digital telephone lines offer faster speeds of 128 Kbps.

In practice these are still too slow and demand has grown for greater bandwidth and faster Internet access. Many companies now offer broadband technology using fibre optic cables offering speeds of at least 2 Mbps (Megabits per second). This technology is sometimes referred to a 'fat-pipe technology.' However these speeds are still slow compared with Local Area Networks that have speeds of 100 Mbps.

ADSL (Asymmetrical Digital Subscriber Line) technology has been developed to give broadband performance using a standard copper telephone cable. Special hardware and efficient compression techniques means that ADSL can operate at speeds of around 2 Mbps. It is called asymmetrical because it can download (or receive data) at speeds of 2 Mbps but upload (or send) data at speeds of 256 Kbps.

As most Internet users receive a lot of data and only send e-mails, ADSL is an attractive option and cheaper than dedicated fibre optic cables. It is not suitable when a user needs to send data at high speeds, e.g. net hosting.

Networks at home?

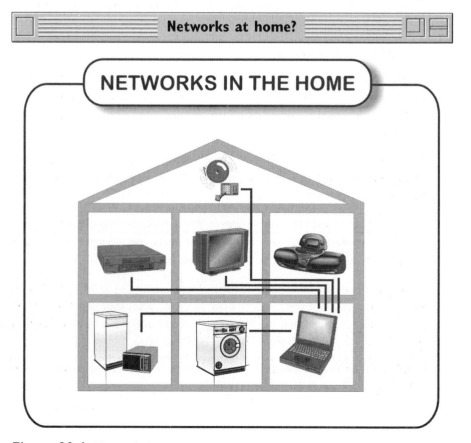

Figure 20.4 Networks in the home

The past decade has seen an enormous rise in the number of PCs in homes. Will networks in homes soon be just as common?

Many homes now do not just have one PC. Some users may also have a laptop for portable computing. Setting up a home network can be just a matter of connecting two PCs using twisted-pair cabling.

The most likely reason for a network would be to let users to share files but it also allows multi-player gaming. In future such a network may be linked to other household appliances such as hi-fis and DVD players. Eventually, the entire entertainment system could be based on a network. You could, for example, program your video recorder using a PC.

Case Study 3

The National Lottery

The National Lottery, run by Camelot, sells tickets in around 35,000 retail outlets. Tills in all the retailers are connected to Camelot's wide area network either by cable or by satellite. As lottery tickets are sold, details of the numbers chosen are entered by optical mark reading (OMR).

The data is transmitted to Camelot's computer centre in Rickmansworth, Hertfordshire. The network needs to be very sophisticated to cope with the large volumes of sales (particularly early in the evening before the draw is made) which have reached over 50,000 transactions a minute. Camelot say that the network has been designed to cope with considerably more traffic than this.

○ What data do you think is transmitted from a lottery network station to the central computer?

○ Give five reasons why Camelot use a WAN for collecting data from shops

Case Study 4

Debit Cards

Over 16 million people in Britain have a debit card, often called a Switch card. These cards can be used in nearly a quarter of a million retail outlets throughout the country. They allow the card holder to pay for goods by having an amount deducted from their bank account. There is therefore no need to use cash in the transaction.

When a card holder presents their card at the till, it is 'swiped' through a terminal. The terminal can check the card against a list of stolen cards. If the purchase is over a certain amount, the terminal connects to a wide area network called SwitchNet to make sure that the card holder has the money in their account and to authorise the transaction. This takes less than five seconds in nearly all cases. Details of the sale are stored in the terminal. Every night the sales data is sent to the bank's computer and the list of stolen cards is updated, using the wide area network.

Each month debit cards are used for over 50 million purchases worth £1.5 billion. Use of debit cards is rising rapidly and mainly replacing the use of cheques.

Summary

- the use of networked computers has expanded as it provides access to more and more information

- networked computers can share data and share hardware

- networked computers give the opportunity for communication e.g. by e-mail

- networked computers offer better security than stand-alone machines

- local area networks cover a small geographical area, while wide area networks cover a large area

Network environment questions

1. Describe two differences between a Local Area Network (LAN) and a Wide Area Network (WAN.) *(4)*

AQA ICT Module 2 January 2001

2. At the central office of a landscape gardening company there are six employees. Each employee has a stand-alone computer system and printer. The company director has commissioned a business survey which indicated that it would be more efficient if the six PCs were formed into a peer-to-peer network.

a) State three benefits that the company would gain from networking their computer systems as a peer-to-peer system rather than a server based system. *(3)*

b) What additional hardware would be needed to connect the six stand-alone computer systems as a peer-to-peer network system? State why each item is required. *(4)*

AQA ICT Module 2 Specimen Paper

3. Explain the function of a gateway when used with Local and Wide Area Networks. *(2)*

NEAB 1998 Paper 2

4. a) Give two differences between a Local Area Network (LAN) and a Wide Area Network (WAN). *(2)*

b) Discuss the relative merits of server-based networks and peer-to-peer networks. *(6)*

NEAB 1997 Paper 2

5. The following are applications which use either a wide area or a local area network or combination of both. For each, justify which network type is most suitable.

a) Cash-dispensing and account-inquiry facility for a national building society. *(2)*

b) Accounting and stock-control system for a department store, using point of sale terminals. *(2)*

AEB Computing Specimen Paper 2

6. A school is investing in 16 computers. They need to choose a between network or 16 stand alone machines. Describe the advantages and disadvantages of each. *(8)*

7. A local surgery uses a number of stand-alone computer systems to manage patient records, appointments, staff pay and all financial accounts. The surgery manager is considering changing to a local area network.

Compare the relative advantages of stand-alone and local area network systems. *(6)*

NEAB 1996 Paper 2

8. In a solicitor's practice there are ten employees working in three offices. Each employee has a stand-alone computer system and there is a shared printer in each office. The head of the practice has been advised that it would be more efficient if the ten computers were formed into a server-based network.

a) State **three** benefits that the practice would gain from networking their computer systems. *(3)*

b) Give **two** reasons for choosing a server-based system rather than a peer-to-peer system. *(2)*

c) State **two** items of hardware that will be needed to connect these ten computers as a server-based network. State why each item is required. *(4)*

AQA ICT Module 2 May 2001

9. The manager of a small hotel uses a stand-alone computer to administer the booking and billing systems. He is considering setting up a small local area network to replace the stand-alone computer, with work stations in reception, in his office and in the dining room.

a) State **two** different types of network that would be suitable. *(2)*

b) i) Give two advantages of this change for the manager. *(2)*

ii) Give two advantages of this change for his customers. *(2)*

c) A friend suggests that a connection to the Internet would also be an advantage to allow customers to book electronically. Explain the extra factors that need to be considered if the customers are allowed to pay a deposit electronically over the Internet. *(4)*

AQA 2000 Paper 2

Human/computer interface

A computer is a tool used by humans to carry out specific tasks. Computers can perform repetitive calculations very fast without error and can manipulate data in ways set up by humans in programs. Modern computers have a huge storage capacity. Humans, on the other hand, have very different strengths. For example humans can easily recognise people and objects and even a small child can interpret language. These are tasks that are hard to achieve using a computer. People are functioning in the real world while computers operate in a digital, electronic environment. Both humans and computers have strengths and limitations – they just happen to be very different. Humans make use of a number of senses to gain information from the world around them including sight, hearing and touch. Computers can input data in an increasing number of ways, and recent developments are producing input devices that more closely resemble the human environment.

Some computer systems run automatically once they have been set up and require no further human involvement. Most such examples are control systems such as a car engine management system. However many systems are interactive, and the design of the place where human and computer meet in terms of hardware device choice as well as the look and feel of the software are crucial to the successful use of the computer system.

Human / Computer Interface (HCI) is a term used to describe the point of interaction between people and computer systems. It is the point where data in human understandable form, for example written or spoken work, images or ideas is converted into digital form for storage and manipulation within the computer or vice versa. When planning an interface a designer has to balance what the user would like and what is achievable in computer terms.

A good HCI will be attractive, easy to use, appropriate to the use for which it is designed and for the users. It will also be safe and robust. The role of the HCI in making the link between human beings and the computer is crucial, as its appropriateness will determine the effectiveness of the dialogue between these two worlds.

An HCI is made up of hardware devices and software. The earliest computers had interfaces that were extremely user-unfriendly and could only be used by people with extensive technical knowledge. As the capabilities of computers have increased, so it has been possible to improve the HCI. Software writers work to make their products both attractive to the user and easy to use. It is quite unnecessary for a user to know in detail what is happening within the computer when he is using it to perform a task. He is much more interested in what it can do in terms of the human world.

Three common forms of communication are *speech*, *written text* and *graphics*. Speech is the most common form of communication between humans but understanding depends on both speakers sharing the same 'natural language'. Systems using communication between humans and computers through speech, though developing, are still limited.

Human/computer communication through written text is commonly used; the speed of communication is limited to the time it takes a user to input text using a keyboard and read the computer responses on the screen. The written word is, like speech, natural language-specific.

The use of graphics is increasingly used in HCIs. As the memory capacity and the processing speeds of computers have increased the use of graphics has become widespread. Icons, tiny pictures designed to convey an easily understood meaning, are fast and easy to interpret and are not language specific.

Sound can also be a feature of an HCI. Audible error messages can alert a user, for example, a printer can 'beep' when a particular error occurs. Such a message would be stored as a sound file. In some cases it is appropriate for software to give audible messages to a user.

HCIs for different environments

There is no such thing as 'the best HCI': different situations and applications have different requirements. The choice of interface will depend upon the physical environment, the experience of the users, the amount of information that needs to be conveyed or gathered and the frequency with which the user will be operating the system. If an application requires large amounts of data to be entered then the most appropriate form of interface might make use of a keyboard.

Users of computer games will know that different types of games require very different kinds of interface. Simple card games can be played using just a keyboard, but are much more fun if a mouse or other pointing device can be used. Good quality graphics in the display enhances the appeal of the game. Other games, such as a flight simulator or adventure game, require high quality moving graphical images, together with excellent sound to be realistic and appealing to the modern player. The use of keyboard cursor keys cannot provide a means of data entry that is sufficiently fast and flexible. Even a mouse or other pointing device is not appropriate – a joystick is needed.

Figure 21.1 Tracker ball

When developing software for use in a primary school to enable young children to develop their maths skills, care must be taken when deciding on the most appropriate interface. A concept key board could be used where pressure sensitive pads are overlaid with different images as required. Screen designs should be very simple with the maximum of visual clues.

A computer system that controls machinery in a factory would require a very different interface. The operator might be working on machinery that requires the use of his hands and the environment might be dirty. The use of voice data entry, based on a simple set of commands, might be the most appropriate in such a situation.

ICT offers many opportunities for people with disabilities, particularly those who have difficulty communicating. There are various computer adaptations available for people who cannot use a mouse or keyboard or who cannot see a normal monitor too well. On-screen keyboards allow users who can use a mouse access to the computer, providing point-and-click access to standard keyboard letters, whole words and communication phrases. Speech recognition means it is not necessary to be able to operate a keyboard or a mouse to use a computer. Output can be to large screens, 'spoken' or in the form of Braille to help users with poor eyesight.

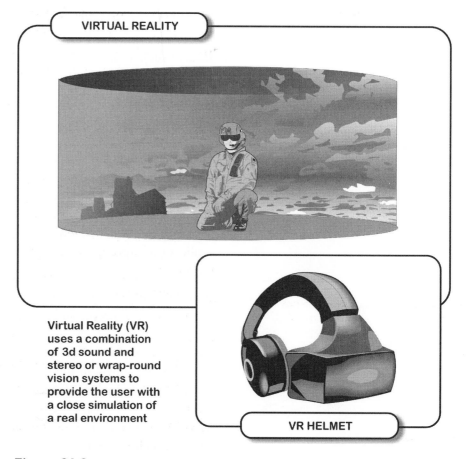

Virtual Reality (VR) uses a combination of 3d sound and stereo or wrap-round vision systems to provide the user with a close simulation of a real environment

Figure 21.2 Virtual reality

Figure 21.3 Railway ticket machine

Railway ticket machines

There are automatic ticket machines at many mainline stations. These machines allow customers to purchase a ticket quickly without having to queue in a ticket office. Customers use a touch screen to choose their destination from a list and the type of ticket, such as single or return. Payment is by credit card: the ticket machine can automatically read the card details from the magnetic strip on the back.

A touch screen is used because

- it is more durable than other pointing devices such as a mouse
- it is easy to operate, even for the inexperienced computer user.

The instructions for use are printed on the screen and are very simple. The user has few decisions to make. The HCI is robust and very easy to use.

Common standard types of human/computer interface

Command line interface

When using a command line interface the user types in commands for the computer to interpret and carry out. The computer responds by displaying text on a monitor screen. The input commands have to be known in advance by the user: there are no clues to help guess them. This type of interface can be used for operating systems, such as MS DOS, and is sometimes used for applications such as data base programs. If a PC running the operating system MS-DOS were to be switched on, a C:> (called the C prompt) appears on the screen. The C represents the computer's hard drive and implies that any file references would be made to that drive. Nothing else happens until the user types in a command. The user has to remember all the commands that she is likely to need. Many commands are complex, with a range of parameters that modify their meaning. Keywords should be chosen which suggest the function and are familiar to the user. Expert users should be able to use abbreviations and default values should be used for missing parameters or prompts.

In MS-DOS the command **DIR** typed in after the prompt results in a list of all files in the directory being displayed on the screen.

DIR/P gives the same list but stops at the bottom of the screen and waits for a key to be pressed before continuing. (the **/P** stands for pause).

DIR/W gives the same list but lists the files in columns on the screen. (**/W** stands for wide)

DIR A: gives a list of files on floppy drive A.

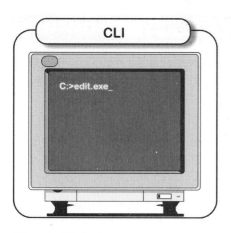

Figure 21.4 Command line interface

Command line interfaces are appropriate in systems used very frequently by experts. Little computer memory is required for such an interface and complex commands can be entered very quickly in one line. The user can give precise sequences of instructions to the computer, allowing complex tasks to be performed. For an occasional or inexperienced user, however, they can be very frustrating as commands will need to be learned or frequent reference will need to be made to manuals. With the first computers, which had very limited memory and processing power, the command line was the only available interface.

Full screen menu interface

A menu-based user interface displays a set of choices on the whole screen so that the user can make a selection from the range of choices offered. Care must be taken when designing the menu system to ensure that there are not too many options on the screen at one time since that can be confusing to the user and lead to mis-selection. To allow for all the possible choices, a hierarchy of sub-menus will need to be devised. The user selects one of the options. This can be achieved either by entering via a keyboard the code letter or number that is displayed next to the option or the menu, or highlighting the option on the screen by moving a mouse or tracker ball, or by pressing the cursor keys. If a touch screen is in use, the user touches the screen at the place where the chosen option appears. When a choice is made another menu may well be displayed. A balance needs to be made between the number of options on a screen at one time and the number of levels of sub-menu required. The more levels needed the longer the system takes to operate and the user is likely to get irritated. A menu-based interface can only deal with situations where the user's requirements are known in advance.

A well designed menu-based interface demonstrates a consistency in menu layout and menu options, using the same prompt for the same operation in all menus and putting the prompt for similar operations in the same position on the screen. Decisions on how to group choices together into sub-menus must be taken carefully by the system designer.

A menu-based interface is appropriate for relatively inexperienced and occasional users of a system. There is little need for learning keystrokes, as the choices are available on screen every time the system is used. The user then only has to recognise the required command from a list. There is much less chance of the user forgetting the command or typing in the wrong command. However, the user is limited to the pre-determined choices that are built into the menus. For frequent users, a menu-based system can prove irksome, particularly if there are a number of levels to pass through every time a choice is to be made.

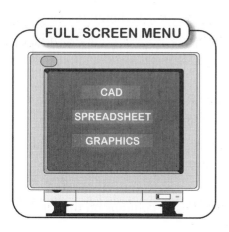

Figure 21.5 Menu-based interface

> **Graphical User Interface (GUI)** >

A Graphical User Interface uses graphics, icons and pointers to make the computer as user-friendly as possible. The user can choose the option required by moving the cursor on the screen to the position required and then clicking or double-clicking on the menu choice or icon. The cursor is moved around by using a mouse, tracker ball or other pointing device. Minimum use has to be made of the keyboard.

A GUI was first developed for the Apple Macintosh using a WIMP (Windows, Icons, Mice and Pointers) environment. Microsoft Windows 98 and Windows 2000 provide GUIs. The GUI is suitable for the inexperienced or infrequent IT user.

Pull
down
menu

Scroll
bar

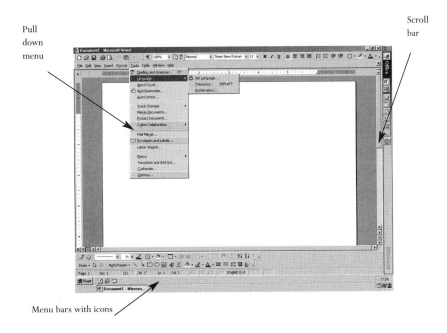

Menu bars with icons

Figure 21.6 GUI

> **Features of GUIs** >

GUIs such as Windows include many features that make dialogue between the user and the computer easier. These include:

Pull-down menus/Pop-up menus

All Windows programs have pull-down menus, making it easy to choose options. With a pull-down menu, the menu title is displayed on a bar. The options are only displayed when the user clicks on the menu title. Pull-down menus can have sub-menus.

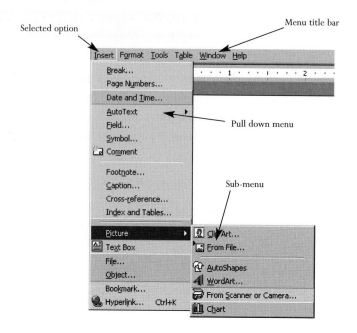

Figure 21.7 Pull-down menu

A pop-up menu is usually displayed when the right mouse button is displayed. The menu is displayed near to the current position of the cursor and is context sensitive:(it relates to the position and object being pointed to).

Icons

An icon is a small picture or symbol which has an easily understood meaning. Icons can be grouped together in toolbars on the screen and can be used as a short cut to a function such as open a file, save, or print. Nearly all Windows software uses icons, which can often be customised to suit the user. Different programs often use the same

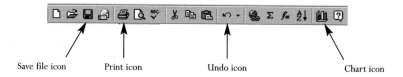

Save file icon   Print icon   Undo icon   Chart icon

Figure 21.8 Icons

icons for common functions. Some users do have problems identifying the icons as they are not always as clear as the designer thought. A new user of Microsoft Word puzzled for a long time over the meaning of the 'tank' icon until it was explained to him that the image was supposed to represent a printer!

Window/dialogue boxes

A window is like a box displayed on the screen that holds the activity of a program. There can be several windows on the screen at the same time. A dialogue box is a window that appears when a particular option has been chosen and more information is needed. For example, when printing a document in Microsoft Word the dialogue box shown in Figure 21.9 appears. The use of dialogue boxes makes it is very easy for the user to be sure that all the questions that need to be answered are answered.

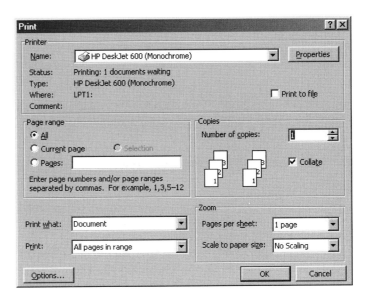

Figure 21.9 Dialogue box

The extra software required for a GUI such as Windows takes up extra disk space. The software is graphics intensive and so needs more computer memory, for example Windows 95 needs at least 80 MB hard disk space and 16 MB RAM to run smoothly. Software using a GUI can be slow to load, take up considerable amounts of RAM and hard disk space as well as require a good quality display on the monitor.

A GUI is likely to have short cut operations such as **hot keys** for the experienced user. A hot key is a special combination of key presses that execute one of the menu commands. For all users, the availability of on-line help, particularly if it is context sensitive, makes it possible to find out how to use new features without the need for extensive training. Context sensitive help allows the user, through pressing a particular key, to obtain more information that relates directly to the operation that is currently being undertaken.

WYSIWYG (What You See Is What You Get – pronounced wizzy-wig)

A very useful feature of modern word-processing and DTP packages is that they provide a screen display that is identical to the version that will be printed. The display exactly reflects layout, font, highlighting such as bold and italic as well as other features. This is known as WYSIWYG.

When using packages that do not support WYSIWYG the user has to insert codes that denote formatting functions such as italics or indentations. Colour or special symbols were used to represent features on the screen, for example, text that had been formatted in italics would appear in red on the screen. Often the only way to see exactly what a document was going to look like was to print it out, which led to many draft copies being produced before a final version was arrived at. Considerable amounts of paper could be wasted.

With WYSIWYG it is possible to see and manipulate the eventual layout of the document on the screen with much greater ease.

Forms based dialogue

Using a form to enter data into a computer system is a very common method. The data entry is structured and, if the form is well designed, the user can enter the data efficiently and with the minimum of errors.

Forms are widely used for data capture, for example when goods are ordered, reservations made, insurance proposals made or questionnaires completed. Forms are used when software is capturing a standard set of data items.

Paper forms are usually filled in from left to right and from the top to bottom; the form filler is able to go back to make changes or even throw the form away and start again with a new one. A form based data entry screen should offer the same facilities. It is very irritating indeed not to be able to edit data that has just been entered on a screen form, and many amateur programs only allow the user to edit data by exiting form filling mode and entering edit mode.

Entries made in a form can either be in text form or can be chosen from a menu or list. The latter has several advantages: it restricts the user to allowable options, reduces the possibility of error and is often faster. Another way of speeding up data entry is to use default settings – the most likely entry is provided and it can be accepted or rejected by pressing a key to position the cursor at the next field on the form.

If more than one screen per form is needed then the form should be split up into logical divisions. For example, a paper booking form for a holiday might require information on customer, holiday destination, mode of travel and car hire all on the same page. This would not fit onto one screen. The form could be split up to have one screen for each of the sections mentioned.

Case Study

Travel Agency

A travel agent uses an information system to help customers chose their holidays. Different types of user use the system in different ways.

Customers can interrogate a local off-line system to find details of all the holidays on offer. Many customers will not have good IT skills. The use of a touch screen might be appropriate so that the user can simply select from a number of choices on the screen to obtain the information they require. The user will not be required to enter any data other than choices. The layout of the screen should be simple and uncluttered to attract the user. The text should be large and good use could be made of colour. There could be a built in printing device so that the customer can print out the details of any holiday which meets their requirements.

Travel agents based in the branches use the system to make bookings on behalf of clients. The system that they used could be built around menus and forms dialogue. Although the agents will be regular users, using the system will only be part of their job, so it will need to be straightforward. Most of the data will be entered using a keyboard, although a mouse or similar pointing device would be used to make menu selections.

ICT specialist staff at the head office set up the system and maintain the accuracy of the database. These users are expert and would spend the majority of their working time using the system. They are likely to be using the system in a variety of ways and will be very knowledgeable about its workings. For these specialists, a command line interface might be the most appropriate as they could carry out complex tasks in the minimum of time.

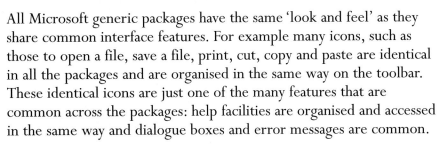

Common user interfaces shared between generic packages

All Microsoft generic packages have the same 'look and feel' as they share common interface features. For example many icons, such as those to open a file, save a file, print, cut, copy and paste are identical in all the packages and are organised in the same way on the toolbar. These identical icons are just one of the many features that are common across the packages: help facilities are organised and accessed in the same way and dialogue boxes and error messages are common.

These similarities make it easy for a user to learn to use a new package, as many of the commands and methods of operation will already be familiar and they will already be able to carry out basic procedures. With this knowledge, they are more likely to have the confidence to investigate the working of other, new features for themselves. The training costs for an organisation will thus be reduced and the productivity of staff will be increased.

Developing new software can be speeded up as sections of code from previous packages, that is tried and tested, can be re-used. From a manufacturer's point of view, a user is more likely to stick to their software once they have become familiar with one package if they are able to transfer much of the expertise that they have built up. Consequently you will find that certain design features become standards that are used by a number of different software manufacturers to take advantage of a user's prior knowledge. An example is the position of the file menu: this can be found on the far left of the main toolbar in a large number of different software packages.

Interface features that are commonly shared

The following are some of the features that have become 'standard' in many packages:

- The layout of the screen – common toolbars are positioned in the same place

- The names given to commands (e.g. 'open' a file rather than 'retrieve' or 'load')

- The shape, colour and design of icons

- The order that the menus appear on the main bar and the way in which options are allocated to menus (e.g. the 'File' menu is found on the left of the bar and the first option in 'File' is usually 'New')

- The way in which the mouse and keyboard can be used to operate menu options

- The cut, copy and paste commands

- Print and print preview commands

- The layout and contents of dialogue boxes

Activity 1

For this activity you need to have access to a software package with which you are unfamiliar. Ideally it should be produced by the same manufacturer as a package which you have used and of which you have a good understanding. For example, you may be a confident user of Microsoft Word who has made some use of Microsoft EXCEL but have never used PowerPoint. Load up the unknown package and explore it. Note down all the features with which you are already familiar – you can include anything here from familiar icons to printing methods. Use the Help facility to find out how to use unknown features of the package.

Case Study 1

The Independent December 2000

The World According to 'Tog'

Bruce Tognazzini reckons that we're all cave dwellers – at least when it comes to using computers. The man who gave us the first Macintosh, with its graphical user interface (GUI), says that things really haven't moved on since 1984.

'The trouble is, you only have a single way to speak to the computer at present, which is the mouse click,' says Tognazzini – better known just as 'Tog'. 'For the average person at home who wants to do some e-mail or finances, the current GUI is adequate. But, once you're using it for serious, professional work, it breaks down. An application such as Photoshop [used by graphics professionals] or Microsoft Office has a screen with hundreds of icons because that's the only way we have to interact with the machine.

'This means that when you want to cut some text or graphics, you have to go to the place on this 'wall' in front of you and point to it and say 'cut', and then point to where you want to move it and find the word that means "paste". It really is like being back in caves where all we can do is go "ugh".

Tog also knows that a lot could be done to improve this: he's seen it done in research projects, at the University of Maryland, for example.

'You could use a gestural interface, so that to copy you would put two fingers on the object on the screen – a screen that would be on your desk, because it's really tiring to hold your hands up to a vertical screen all day – and then pull them apart... now, that would be a big improvement. It would mean that we had more words in our vocabulary for interacting with the computer. And you could mix those in with gestures and voice recognition.'

How we interact with computers is Tog's great passion. Too little attention is paid to it, he believes. He's frustrated, too, with his former employer. Apple could still make our interaction with computers better, but for some reason won't. He says it holds patents on a method of grouping documents called 'piles', developed nearly 10 years ago by a group led by Gitta Salomon.

Describing it on his Web page (www.asktog.com), he said: 'A pile is a loose grouping of documents. Its visual representation is an overlay of all the documents within the pile, one on top of the other, rotated to varying degrees. In other words, a pile on the desktop looked just like a pile on your real desktop.

'To view the documents within the pile, you clicked on the top of the pile and drew the mouse up the screen. As you did so, one document after another would appear as a thumbnail next to the pile. When you found the one you were looking for, you would release the mouse and the current document would open. Piles, unlike today's folders, gave you a lot of hints as to their contents. You could judge the number of documents in the pile by its height. You could judge its composition very rapidly by pulling through it.'

- Describe, in your own words, the limitations that 'Tog' feels exists with current GUIs.

- Summarise the improvements he describes.

Activity 2

For this activity you will need access to some amateur software. Shareware software is a possible source or perhaps projects produced by past students!

Run the software and study the HCI very carefully. Referring to the section '**Rules for good interface design**' above, pick out five good features and five bad features.

Activity 3

Design an interface for each of the following systems. For each system specify hardware devices, type of interface (such as command line, menu-based, GUI) as well as designing any necessary screens.

- A simple painting program for young children

- An information system for use by tourists that gives information of a town's highlights

- A theatre booking system

Rules for good interface design

Clarity of structure and layout is vital for an effective layout. Features such as headings, error messages and request for user responses should be placed in a consistent location. Messages to the user should be separated from the body of the screen.

When using menus, the heading for each menu should be easily distinguished from the choices. Just as the choices should be easily distinguished from each other, there should be adequate blank space between the choices to make them clear.

When designing an interface for an inexperienced user a number of considerations should be taken into account. Messages should be clear, unambiguous and friendly, the language simple to understand and jargon free. Inexperienced users are likely to make mistakes so the software should confirm all requests for action from the user (for example are you sure Y/N).

All system interfaces should be consistent and should look, act and feel the same throughout. The level of complexity should be appropriate to the experience of the expected user. Jargon should be avoided and different names should not be used for the same thing in different parts of the package, for example save, keep, store, write; invalid, not valid, illegal, not legal. Care should be taken that the amount of explanation given and the information load is at the right level for the user. If it is too high then the user is likely to become confused and make errors. If it is too low then the user could become bored and make errors due to lack of attantion.

Natural language interface

In the real world humans communicate in natural language such as English. A natural language interface with a computer, where requests for data and commands could be typed in normal English or, even better, spoken would be the ultimate interface in most situations. Natural language is extremely flexible and powerful. However, to easily interpret language a computer needs it to be precise with a limited number of commands and strict syntax rules. English is less precise with a much greater vocabulary. In fact new words appear and the meaning of other words changes over time. Sentences can be ambiguous and it is only when the context is taken into account that the meaning is clear.

It is not yet possible to develop systems that accept a natural language but some systems, mainly expert systems, have been developed using a limited sub-set of a natural language, that is by restricting instructions to a number of acceptable commands. The user still has to learn how to use these commands and how to construct sentences that can be understood by the system. This may defeat the object of trying to develop an easy-to-use natural language.

Commands are usually typed in at a keyboard, but if such a system is combined with speech recognition, there is potential for humans and computers to have a limited dialogue.

Pure natural language interfaces may still be the stuff of science fiction but there are many research projects exploring the possibilities of talking to computers. The Spoken Language Systems (SLS) Group at MIT's (Massachusetts Institute of Technology) Laboratory of Computer Science have been trying to get computers to communicate the same way as humans since 1989. Progress has been made as can be seen from the following excerpt from a conversation between a user and JUPITER, an SLS-based weather forecast system.

> User: *Yes, I would like the weather forecast for London, England, please.*
>
> JUPITER: *In London in England Wednesday, partly cloudy skies with periods of sunshine. High 82 and low 63. Is there something else?*
>
> User: *What is that in degrees Celsius, please?*
>
> JUPITER: *In London in England Wednesday, high 28 Celsius and low 17 Celsius. What else?*

Check out their progress at their web site:
http://www.sls.lcs.mit.edu/sls/whatwedo/index.html

The search engine 'Ask Jeeves' allows the user to ask questions in natural language. It builds up a bank of questions and uses these to help interpret new ones.

Quote from Ask Jeeves:

'From the beginning, Ask Jeeves at Ask.com has allowed users the ease of interacting on the Web the same way they do offline. Users pose questions in plain English and receive links to Web sites containing relevant information, services and products. Ask Jeeves' combination of natural language question answering, human editorial judgment and popularity technology gives users the benefit of millions of previous searches. Growing smarter with each interaction, Ask Jeeves provides targeted, relevant responses to user queries. This simple, straightforward approach has made Ask Jeeves sites some of the most highly visited sites on the Internet.'

Figure 21.10 Ask Jeeves

Ambiguities in language

Languages like English can be ambiguous. Some words have multiple meanings and we often only know the specific meaning from the context of the sentence. For example, the word *lead* has several meanings. It can mean the leash for a dog. It can mean the person in front in a race. Pronounced differently it can mean the writing part of a pencil. The written sentence: '*I want the lead*' could mean:

a) *I want the leash for my dog*

b) *I want the lead to put in my pencil*

c) *I want to be in front*

There are probably many other meanings too.

The word 'by' is interpreted in different ways in different contexts:

> *'The lost children were found by the searchers' ... (who)*
> *'The lost children were found by the mountain' ... (where)*
> *'The lost children were found by nightfall' .... (when)*

The structure of a sentence can also be ambiguous. If you were to say: 'My car needs oiling badly' would you really want someone to make a bad job of oiling your car? What do you make of the sentence: 'Fruit flies like a banana'?

Summary

Computers are very good at repetitive calculations and manipulation of data, but they are only useful in situations for which they have been programmed

Common interfaces are

- command line Interface (CLI)
- full screen menu
- graphical User Interface (GUI)

The HCI must be designed specifically for a given environment – different situations and users at different levels need very different interfaces

It is not yet possible to program computers in 'natural languages' like English as computer commands need to be very precise and English has a large vocabulary

Advantages of a natural language Interface	Limitations of a natural language interface
It is the natural language of humans, who can express themselves freely without constraint	Natural language is ambiguous and imprecise
No need for special training	Natural language is always changing
Extremely flexible	The same word can have different meanings

Human/computer interface questions

1. A manufacturing company intends to use an information system to store details of its products and sales. The information system must be capable of presenting the stored information in a variety of ways. Explain, using three distinct examples, why this capability is needed. *(6)*

   NEAB 1996 Paper 2

2. Most modern PCs make use of a GUI (graphical user interface) and have a WYSIWYG (What You See Is What You Get) word-processing package.

   a) State **four** characteristics of a GUI. *(4)*

   b) Describe three advantages to the user of a WYSIWYG word-processing package. *(6)*

   AQA ICT Module 2 Jun 2001

3. All the staff in a small office use the same word-processing, spreadsheet and database packages. These packages all have a common user interface.

   a) Give four advantages of having a common user interface. *(4)*

   b) State four specific features of a user interface which would benefit from being common between the packages. *(4)*

   AQA ICT Module 2 Jan 2001

4. A travel agent uses an information system to help customers chose their holidays. The system is used by different types of user. Justify different user-interface features which would be appropriate for each of the following:

   a) customers, who can interrogate a local off-line system to find details of all the holidays on offer

   b) travel agents, who use the system to make bookings

   c) staff who set up the system and maintain the accuracy of the database. *(10)*

   NEAB 1996 Paper 2

5. A different human-computer interface would be needed for each of the following users:

   a) a young child in a primary school

   b) a blind person

   c) a graphic artist
      i) For each user describe and justify an appropriate human-machine interface. *(9)*

   NEAB Specimen Paper 2

6. User-interfaces have gradually become more and more oriented to the needs of users over recent years.

   a) Briefly describe three features of user interfaces which have been developed and explain how each has benefited the user.

   b) Describe two ways in which user interfaces could be further developed to make computers more accessible and friendly to untrained users.

   London Board Computing Specimen Paper 1 q9

7. Give three reasons why you think speech recognition is likely to expand in use.

8. Many machines now offer a graphical user interface such as Windows.

   a) Describe two features of such interfaces that are likely to be helpful to a non-technically minded user.

   b) Give three disadvantages of this type of interface.

   AEB Computing Specimen Paper 2

9. A college uses a range of software packages from different suppliers. Each package has a different user interface. The college is considering changing its software to one supplier and to a common user interface.

   a) Give four advantages of having a common user interface. *(4)*

   b) Describe four specific features of a user interface which would benefit from being common between packages. *(4)*

   c) Discuss the issues involved, apart from user interfaces, in the college changing or upgrading software packages. *(8)*

   NEAB 1997 Paper 2

10. Describe and justify a suitable human-computer interface for:

a) a user of a bank ATM (Automatic Teller Machine)

b) a games computer

c) a teletext user

d) a computer programmer.

11. A large entertainment and leisure complex has a wide range of facilities available including a cinema, live entertainment, indoor sports and exhibition facilities. They are considering introducing a computer based information system to be used by the general public who visit the complex to find out details of future events.

a) Using an example describe how a user might interrogate the computer. *(2)*

b) Describe a human/computer interface suitable for this system. *(4)*

c) Describe the hardware required. *(4)*

d) What steps must the complex owners take to ensure the timeliness and accuracy of the information displayed.

Based on NEAB 1998 Paper 2

12. Many schools and colleges use local area networks of personal computers to allow their students to access packages and to store their files. Most students word process their assignments and use packages to assist their learning. Some students also learn to use database packages, and others write large programs for project work.

a) List three of the main functions performed by a network operating system.

b) After logging onto the network, students are presented with a personalised graphical user interface (GUI).
   i) State **two main features of a** GUI.
   ii) Explain why a GUI is preferable to a command line interface.
   iii) Explain a benefit for the students of having a personalised GUI.

c) The college needs to upgrade the version of the word processing package currently in use. However the network manager asks the management to wait until the end of the academic year. State two reasons why this is a sensible request.

NEAB 2000 Paper 2

Glossary

Alphanumeric characters	Letter, numbers or other character, for example punctuation marks
ASCII	(American Standard Code for Information Interchange). The binary code used in computers to store alphanumeric characters.
ATM	(Automatic Teller Machine). The Official name for cash machines outside banks.
Back-up	To make an extra copy of stored data in case the original is lost or corrupted
Bandwidth	physical limitations of a communication system (usually bits/sec)
Batch processing	A form of processing where all the information is batched together before being processed
Bit	(Binary digit). A binary number which can only have the value 0 or 1.
Bitmap	An image which stores the colour of every pixel.
Broadband	A data transmission method that involves several channels of data and so is faster than older methods
Browser	A program that allows the user to access a database (typically the Internet)
Buffer	Memory where data is stored while waiting to processed, typically in a printer
Bugs	Errors in computer programs
Byte	A group of eight bits, normally storing one alphanumeric character
Cache	A very fast but more expensive computer memory
Caching	Storing Internet files locally – usually on the computer's hard drive – to enable the files to load quickly if revisited.
CAD	Computer Aided Design
CD-ROM	(Compact Disc-Read Only Memory). A small plastic disc used to store data.
Compression	A method of reducing the size of a file, typically to use less disk space
Configure	To set up a computer system for the appropriate hardware and software. A system will need to be configured for the printers, sound cards and so on.

Cyber-   A prefix alluding to computer communication often with reference to the Internet as in cybershopping, shopping by computer, cyberspace, everything accessible by computer communications.

Database   A structured set of data stored on a computer

Data integrity   The reliability of data, that is ensuring it is accurate

Data security   Keeping data safe from loss

DBMS   A set of programs allowing the user to access data in a database

DDE   (Dynamic Data Exchange). Shared data in two packages is linked so that when it is updated in one program, it is automatically updated in the other program.

Debug   Remove bugs from a program

Desktop   An icon-based user interface that enables the user to load software easily. When you load Microsoft Windows, you see the desktop.

Digital   Something that is represented in numerical form typically in binary numbers

Direct-mail   Advertising a product by sending details directly to potential buyers through the post

Directory   An area (usually of a disk) where files are stored. A disk may have several directories and sub-directories to make finding files easier and to aid security

Dongle   A piece of hardware, for example a lead that has to be plugged in to the computer before software will run. Usually used to protect copyright

DOS   Disk Operating System

Dot.com   A company usually trading exclusively via the Internet

dpi   Dots per inch – describes the performance of a printer

e-banking   The use of the Internet to communicate with your bank.

e-commerce   The use of computers and electronic communications in business transactions, including web-sites, EDI, on-line databases and EFTPOS systems.

EDI   Electronic Data Interchange. Transferring information such as orders and invoices electronically between tow organisations.

e-tailors   Retailers who do business on the Internet.

e-shopping   Using the Internet to purchase goods and services.

EFTPOS   Electronic Funds Transfer at Point of Sale. The system where customers can pay by debit (Switch) card and the money is taken electronically from their bank account.

Embedding   Including one file (such as an image or a document) in another file. See OLE (Object Linking and Embedding).

Encryption   To scramble data into a secure code to prevent it being read by unauthorised users

Extranet	The linking of two intranets usually to assist business transaction, for example linking a customer and a supplier
FAQ	Frequently Asked Questions. A file containing answers to common questions, for example about using a program.
Fax modem	A modem that enables a computer to send and receive faxes
Fibre optic	A cable made out of glass fibre and used in communications
Filters	An option in a program enabling the user to import files from or export files to another program
Flatbed scanner	A scanner in which the item to be scanned is placed on a flat piece of glass
Floppy disk	A small removable disk in a hard plastic case, used to store data
Gigabyte (GB)	A measure of memory capacity equal to roughly 1 000 000 000 bytes (it is exactly 2 to the power 30 or 1 073 741 824)
GUI	Graphical User Interface, for example Windows. It is sometimes pronounced 'gooey'.
Hacking	Unauthorised access to a computer system, possibly for criminal purposes
Hand scanner	A small device, held in the hand and dragged over the item to be scanned
Hard disk	A magnetic disk inside a computer that can store much more data than a floppy disk. Usually it cannot be removed but removable hard disks are becoming more common.
Hardware	The physical parts of the computer, such as the processor, keyboard and printer
HTTP (HyperText Tranfer Protocol)	The standard protocol for sending and receiving data on the Internet
Integrated package	A package which combines several different applications such as a word-processor, a graphics package, database, communications software and spreadsheet
Interactive	A system where there is communication between the user and the computer
Internet	An international WAN providing information pages and e-mail facilities for millions of users
Intrenet	A private internal network using Internet software, that can be used for internal e-mail and information
IRC	Internet Relay Chat. A function of the Internet allowing users to send and receive real-time text messages.
ISDN	Integrated Services Digital Network. A telecommunications digital network which is faster than an analogue network using a modem.

ISP	Internet Service Provider. A company that offers a connection to the Internet.
Java	A programming language used for utilities on web pages
JPG or JPEG	Joint Photographic Expert Group. An ISO standard for storing images in compressed form. Pronounced jay-peg.
Kilobyte (KB)	A measure of memory capacity equal to 1024 bytes
Licence agreement	The document which defines how software can be used, particularly how many people can use it
Macro	A small program routine usually defined by the user
Magnetic disk	A small disk coated with magnetic material on which data is stored. It can be a floppy disk or a hard disk.
Magnetic tape	A long plastic tape coated with magnetic material on which data is stored
Mail-merge	A feature of a word-processing program that combines details from a file of names and addresses into personal letters
Master file	The file where the master data is stored. Data from this file is combined with data from the transaction file.
Megabyte (MB)	A measure of memory capacity equal to 1 000 000 bytes (it is exactly 2 to the power 20 or 1 048 576)
MICR	Magnetic Ink Character recognition. The input method used to read cheques.
Modem	Modulator/demodulator. The device that converts digital computer data into a form that can be sent over the telephone network.
MS-DOS	Microsoft Disk Operating System. The operating system developed for the PC.
Multi-access	A computer system allowing more than one user to access the system at the same time
Multimedia	A computer system combining text, graphics, sound and video, typically using data stored on CD-ROM
Multi-tasking	A computer system that can run more than one program simultaneously
Network	A number of computers connected together
OLE	(Object Linking and Embedding). A method of taking data from one file (the source file) and placing it in another file (the destination file). Linked data is stored in the source file and updated if you modify the source file. On the other hand, embedded files are part of the destination file.

On-line Processing	Processing while the user is in contact with the computer
Operating system	The software that controls the hardware of a computer
Package	A program or programs for a specific purpose
Palmtop	A small handheld computer around the size of a pocket calculator
Peer-to-peer	A type of network where there is no server, with each station sharing the tasks
Pentium™	A processor developed by the Intel Corporation™ for the PC
Peripheral	Any hardware item that is connected to a computer such as printers, mice or keyboards
PIN	Personal Information Number, used to check that the user is the person they claim to be, for example at an ATM
Platform	Used to describe a hardware or software environment
Port	A socket usually at the back of the computer.
Portability	The ability to use software, hardware or datafiles on different systems
Primary Key	A unique identifier in a record in a database
Protocol	A set of rules for communication between different devices
QBE	Query By Example. Simple language used to search a database
RAM	Random Access Memory. The computer's internal memory used to store the program and data in use. The contents are lost when the power is turned off.
Redundant data	Data that is repeated unnecessarily (in database)
ROM	Read Only Memory. Part of the computer's memory that is retained even when the power is turned off. Used to store start up program and settings.
Serial access	Accessing data items one after the other until the required one is found. Associated with magnetic tape.
Server	A dedicated computer that controls a network
Shareware	Software that can legally be distributed freely but users are expected to register with and pay a fee to the copyright holder.
Smart card	A plastic card, like a credit card, with an embedded microchip. The information in the chip can be up-dated, for example when cash has been with drawn from a ATM.
Software copyright	Laws restricting copying of software
Software	A computer program or programs

Systems analyst	A person whose job involves analysis whether a task could be carried out more efficiently be computer
Toggle switch	A switch or button which if pressed once turns a feature on. If pressed again it turns the feature off. The Caps Lock button is an example.
Transaction file	A file containing new transaction details or changes to old data, which is merged with the master file
USB	Universal Serial Bus – a port on the back of a computer used to connect peripherals such as scanners or a palmtop.
USB hub	A device that plugs into the USB port that enables several peripherals to connect to the computer at once.
URL	Uniform Resources Locator – the Internet address, e.g. www.hodderheadline.co.uk
Vector graphics	Image system that stores lines by the length and direction rather than the individuals pixels (as in a bit map)
WIMP	Windows, Icon, Mouse, Pointer.
Windows™	A GUI for the PC produced by Microsoft
Wireless network	A network that uses radio waves to transmit data rather than cables
WWW	The World-Wide Web
WYSIWYG	What You See Is What You Get

Answers

These answers are intended to be brief. Exam answers should be in sentences.

 Chapter 1

1. Garbage in, garbage out. If the data input into a computer is wrong, the information coming out will be wrong too. See chapter for examples.

2. ICT means using any form of modern technology for the collection, storage, processing and sending of information.

3. (a) **Input** entails entering raw data into the computer

   Processing entails converting raw data into a form which is useful to the user.

   Output entails transferring processed information to the people who will use it

   Feedback: is output that is returned to the user to help them refine the input phase (must have both points).

   (b) Information is meaningful data in a useful form whilst knowledge is applying information to make decisions.

4. If a value judgement is coded, there may not be a code that fits your view. This can lead to coarser or inaccurate data. e.g.

   This new product will cost £1.65. Is it:

   ☐ much too cheap
   ☐ too cheap
   ☐ about right
   ☐ too expensive
   ☐ much too expensive

The person questioned may think it is between too expensive and much too expensive.

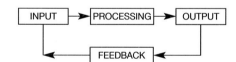

5. See answer to question 3.

Input – entering a bar code at a supermarket till
Processing – looking up the product name and price and calculating the total bill
Output – displaying the product, price and total bill

6. (i) is not true as some customers may think the service is poor
(ii) is true
(iii) may not be true as the average does not say anything about consistent performance

7. There will normally be a limited number of codes for value judgements. This forces coarseness and reduces its accuracy, e.g. blue/green eyes must be coded as blue or green. Loss of meaning depends on the weight given to it by the coder of the information.

8. See text.

9. (a) The speed at which data can be processed allows quick turnaround and therefore more accurate, up to date results which is important when consumer demand changes quickly. Data can be manipulated more easily so more statistical techniques and graphing can be used. It can handle a greater volume of raw data so larger samples can be taken – leading to more accurate results.

(b) **Data**: raw facts: figures or measurements with no processing having been carried out – no use to anyone – for example one questionnaire by self has no meaning.

Information: data that has been processed to give it meaning, for example if the total number of responses to a given question is known, a graph can be drawn of the results.

Chapter 2

1. Three factors that affect value could be accuracy, up-to-dateness, and relevance. Sales figures need to be accurate and up-to-date. They need to be relevant to that product.

2. (a) Four items of data which are captured:
account number/employee no/credit card no
amount of sale
date and time of sale
name of shop/location

(b) Information derived from this information.
Tracking where the sales representative has been
What has been bought, e.g. petrol is likely to have been bought at a petrol station
Analysis of spending patterns
The distance travelled and time taken between stops

3. (a) See question 1

  (b) labour costs in collecting and organising data
    hardware costs
    cost of transmitting, processing, storing data

4. See question 1

5. Poor information. May over-stock. Too much money tied up in stock, could go off. May under stock. May run out. This will lose sales and customers will be dissatisfied.

6. (a) e.g. Account no – data. Name and address of customer on envelope – information.

  (b) Is it up-to-date? If the data has not been updated today, information on how much the customer owes may be wrong. Is it complete? Have details of all the customer's purchases been entered. If not information from the system is not accurate.

7. (a) Fashions may change; people may have more money and go to different places. Out-of-date information may not be accurate.

  (b) Company may have too many/few holidays available – will lose money either way.

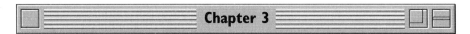
Chapter 3

1. (a) Databases store details of customers. Mail merge sends personally addressed targeted letters to selected people in the database.

  (b) Information on customers' age, income, past purchase, interests, etc. may be stored. From this the company can build up a profile of each customer and whether they are likely to be interested in the product.

2. A more likely explanation is that the wrong data has been entered into the computer. For example an electricity meter has been read incorrectly or the wrong number entered. Or it may be that the system has not been tested fully and the wrong calculation has been made.

3. The letters are all personally addressed and so more appealing to the customer than the impersonal Dear Sir. Letters can be targeted to those people most likely to buy.

4. Reasons include: quicker for the computer to perform calculations than manually, computers are more accurate than humans, paying in cash is time consuming to count, paying in cash has security risks, payroll programs can be run overnight when other computer operations are not taking place – so there is spare capacity on the system.

Chapter 4

1. Credit or debit cards; standing orders and direct debits; automatic payment of salary.

2. No: cash likely to be needed for small purchases: market, pocket money etc; not everyone could have device to input payments; consumer resistance. Yes: more and more purchases are electronic; could use smart cards.

3. (a) Benefit: make better use of doctor's time; danger: the computer may not be programmed to recognise every disease.

   (b) Benefit: pupils can have access to a greater range of subjects and knowledge; danger: loss of personal interaction

   (c) Benefit: consistency; danger: would not take account of special circumstances

4. Social Impacts: **telecommuting** brings ability to work from home for an increasing number of people; computer use has brought new types of **crime** which needs new legislation; **blurring of boundaries** between homework etc.

   Organisational impacts: **dependency on computers** – organisations can lose millions of pounds if computer breaks sown; regular **training** is necessary to maintain a work force with up-to-date skills.

5. (a) ordering can be done very quickly; store can be kept fully stocked and there will be a reduction in staff time spent on stock ordering

   (b) a customer is more likely to find goods they require in stock if levels maintained automatically

6. (a) Distance learning allows individuals to take courses that would otherwise be unavailable; the use of Internet provides the opportunity to research topics

   (b) Computer based records make it easier for staff to find information on patients; Monitoring equipment provides instant information on the condition of patients and can cut staff costs; expert systems provide an aid to diagnosis

   (c) On-line shopping saves time and enables the housebound to shop; e-mail allows easy communication with friends and family

   (d) Electronic communication allows employees to keep in touch with customers and other outside agencies more easily

   (e) JIT (Just in time) manufacturing saves on storage space and costs and avoids waste; robots reduce staff costs

   (f) Databases allow easy search and retrieval of data and help solve crimes faster

Chapter 5

1. (a) e.g. mobile phones and satellite TVs

   (b) When you send an email it is stored in a 'mailbox' on the computer of the recipient's Internet Service Provider until the recipient logs in later to retrieve their mail.

2. (a) Videoconferencing means communicating between two or more places so that the correspondents can both can and hear the person they are talking to. They are likely to use computers and high speed cables to send images from one computer to the other.

   (b) Advantages: No need to travel so saves time and money, can share computer files – i.e. both can work on the file at the same time. Disadvantages: time differences between the UK and Australia means that one company is operating at night, set up costs are expensive, they couldn't demonstrate the quality of a building material very easily.

3. (a) e.g. Address books, groups of addresses, reply function, forward message, send attachments, encrypt messages

   (b) e.g. provides hard copy, arrives almost immediately, can receive even if you are doing something else, can set up groups of addresses and send at once.

   (c) Proper staff training in the use of e-mail to ensure that staff realise implications of their actions; set up procedures for the removal of old e-mail to prevent excessive storage; limit staff disk space to limit amount that they can store e-mails and only send them when there are several to be sent (e.g. once an hour), prohibit large file attachments…

4. **Advantages**: Can access e-mail wherever they go (unlike snail mail), e-mail arrives almost immediately, they can read them when they want to (unlike phone) **Disadvantage**: the salespersons may not check their mailbox regularly.

5. No need to type in names – already on school database. Fewer mistakes. Arrives quickly at exam board. Board do not need to type them in on their computer. School can get results by e-mail and import them into database software…

6. (a) Describe reply and address book

   (b) see chapter

   (c) Internet access, send attachments, shared diaries

7. See text.

8. (a) EDI is Electronic Data Interchange – companies being able to contact each other electronically and share electronic data. For example, this might include orders, invoices, prices, production levels between supplier and customer.

(b) Advantages of EDI include faster transfer of data, reduction of errors, saving of time in producing these documents by hand on paper, cutting down on paperwork, ability to make decisions more quickly because more information is available…

9. (a) Describe how to forward a document

(b) Describe how to set up and use a group of names in an address book

(c) Describe how to attach a document

(d) Describe how to increase priority of an e-mail

10. (a) Modem to convert digital signals to go down the phone line. Phone line to connect to the Internet.

(b) (i) Browser looks at web pages
(ii) Editor sets up web pages
(iii) Email software sends and receives e-mails

(c) Benefits include ability to search through millions of pages of information. Can send several e-mails at once. Access is for the price or a local call or free. Can send messages even if they are not in the office …

Chapter 6

1. Good inter-personal and verbal communications skills needed to explain to a user what to do to use the system;

Good written communications skills needed to write a manual or report for management; willingness to work flexible hours (computer systems may run over night).

2. Good inter-personal and verbal communications skills, the ability to work well in a team, a flexible approach; an analytical approach to problem solving.

3./4. Willing to work flexible hours – user support roles require the ability to stick at problems and see them through; be able to communicate well orally to enable efficient and effective communication with users or colleagues e.g. interviewing and questioning effectively to obtain end user requirements; good written communication skills – ability to write documentation both technical and end users; get on well with a wide variety of people – ability to work as part of a team means being able to exchange views, share information, is the usual way of working in many IT establishments.

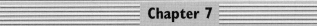

Chapter 7

1. Back-up files and store in a secure place offsite; use keyboard locks to prevent unauthorised people from using computer terminals; take precautions against viruses (see chapter for more detail).

2. See chapter.

3. See chapter.

4. Stock manager should have read/write access as they should have the ability to add, delete and amend records of stock and change prices. Sales staff should have read/write access as they need to be able to see details about stock and to change data as sales are made. Store manager should have read only access as he needs to be able to view the data but not change it.

5. Students seeing inappropriate material such as pornography could be prevented by the use of software that denies access to inappropriate sites. Viruses being downloaded; could be prevented by use of anti-virus software which would detect any such threats.

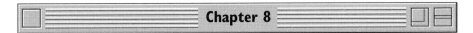

Chapter 8

1. The three types of offence are unauthorised access to computer material, unauthorised access with intent to commit or facilitate commission of further offences and unauthorised modification of computer material. Examples are in the text.

2. See question 1

3. Set up FAST, prosecuted offenders, used dongles.

4. (a) The licensing agreement is made between user and producer and specifies exactly how the software can be used.

   (b) This depends on the wording of the agreement and Mr Patel but Mr Patel is probably breaking the terms of the licence. It is likely that even if Mr Patel uses the software on only one machine at a time, the software is <u>installed</u> on both machines at once. While Mr Patel is using his laptop, a colleague at work may be using the same software on Mr Patel's PC at work and so breaching software copyright law.

   (c) A network licence or a site licence for a minimum of 20 copies.

   (d) See question 1.

Chapter 9

1. (a) see chapter.

   (b) national security, prevention and detection of crime, collection of taxes, family or household affairs.

   (c) disclose data on an individual to the person concerned, correct any mistakes.

2. (a) Print out current details every year; give to students, they amend with any changes and hand it back to ensure that computer details corrected.

   (b) Will the data be passed on to a third party? How long will the data be kept? Will the data be used for direct sales (e.g. by telephone.)

3. See chapter.

4. (a) The Data Protection Act requires it.

   (b) e.g. name and address, description of data, purpose, from whom the information was obtained, to whom the information will be disclosed, countries where the data may be transferred.

5. (a) Easy to store names and addresses in a database. Easy to merge details in a mail-merge letter.

   (b) Keep records of past purchases. Send advertisements for these products only – they are likely to buy again.

   (c) We are registered with the Information Commissioner, we delete all personal information that we no longer use, you can inspect details stored on you if you want, you could have ticked a box on a form if you did not want to receive this sort of mail...

6. It is unlikely that the Data Protection Act has been broken. It depends on the purpose for which the information is registered to be used.

7. They would appear to have broken the second principle, i.e. personal data has not been used for the specified purposes.

Chapter 10

1. Colours that are too bright can cause eye strain; continuous annoying sound can cause headaches; flashing screens can induce an epileptic fit.

2. Tilt and turn screen prevents back or neck-ache; low radiation monitor prevents eye-strain; an ergonomic keyboard helps to prevent RSI.

3. Non adjustable keyboard can cause RSI; a poor sitting position can lead to backache; the reflection and glare from screen can cause eye-strain.

4. RSI stands for repetitive strain injury which affects the shoulders, fingers and the wrists. Symptoms are stiffness, pain and swelling.

5. See question 2.

6. Bad posture; flickering screens; lack of breaks; radiation from monitors; repeated use of the keyboard; ozone emissions from laser printers.

9. See answers for 10.

10. (a) e.g. back-strain, eye-strain and RSI.

    (b) adjustable chairs (back-strain), tilt and turn monitors (back-strain), screen filters (eye-strain), ergonomic keyboards (RSI), regular breaks (all three) (answers <u>must</u> be linked to answers in 10a)

11. See question 10.

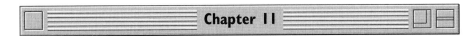

Chapter 11

1. OMR: reading questionnaires; Bar code scanning: identifying goods; MICR: sorting cheques.

2. Improved software; reducing costs; increased memory of computers.

3. Bar code scanning: cheap but hard to read as many runners past post; marshall keys in number that is displayed on runner into laptop: cheap and easy but could mistype; runners use swipe card: may have to queue up to use.

4. OCR:
   High volume justifies cost.
   Use data validation (see Chapter 12).

5. (a) Quicker to enter than typing for non-specialist typists; leaves hands free for other tasks.

   (b) Giving commands e.g. in a factory using a computer controlled machine; entering text such as a letter.

6. Unclear speech; background noise; insufficient training on new voice.

7. (a) Accurate input: there is an audible signal if bar code misread and the use of a check digit to ensure bar code read correctly; ease of changing prices without re-pricing stock.

   (b) Prices not visible on product so customer needs to look elsewhere e.g. on shelf for price.

8. (a) Typed text and handwriting needs to be clear otherwise conversion to characters may not take place; damaged or folded documents affects accuracy of results; as results are not always accurate the imported text needs to be proof-read/spell-checked.

(b) Scanning a document is faster than typing it so there is a reduction in staff time required; text to be stored in a form that can be edited allows updating of previously produced documents without completely retyping them.

(c) Maps, plans, pictures or photographs, signatures.

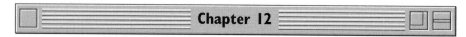

Chapter 12

1. (a) Total length is 7 characters. No lower case characters. 3rd and 4th characters are numbers. Last 8 characters are letters..

   (b) The number might be a valid number, but not the actual registration number of the car in question.

2. Last 2 digits between 00 and 99; middle two digits between 01 and 12; 1st 2 digits between 01 and 31, 30 or 28 depending on value of middle 2 digits.

3. See chapter 10.

4. (a) Check digit on candidate number; candidates name made up of alphabetic characters; subject one of an allowable list in look-up table

   (b) A valid subject could be entered, but the candidate might not be entering that subject.

5. (a) Accurate data is correct, valid data is feasible. 24051967 is entered as my date of birth. It is valid as it represents a date: 24th May 1967 but it is not accurate as that is not my birthday!

   (b) Mistyping and copying (transcription errors) can be avoided by double entry – data verification.

6. (a) Data can be keyed in – DDE. This requires no specialist equipment but prone to transcription error. Forms could be read by OCR which removes the need for transcription. However, some handwriting could be hard to read and forms could be damaged.

   (b) Use of computer software to detect data that is incomplete or unreasonable. Can use check digit for catalogue number or Agent Id code; range check on quantities, presence check on agent code, name, quantity, lookup on catalogue number.

7. (a) Number of tickets, venue of festival, name of festival, date of festival, type of ticket, price of ticket, method of payment.

   (b) **Number of tickets**: range check, format check, presence check or type check; **venue**: format check, presence check, look-up list; **date**: format check, presence check, range check, cross field check with venue.

8. (a) The computer would print the address and name. The account number, previous account number and expected high reading.

(b) Range check on meter reading to check that current reading >= previous reading; format check on meter reading to check handwriting has been read as all digits.

(c) Because the meter reading may still be incorrect even if they are valid. This will happen if the meter reading has been written down wrongly or incorrectly read by OCR.

(d) See Chapter 16.

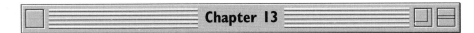

Chapter 13

1. See chapter

2. Product number, Product name, Product Description, Price, Number in Stock, Re-order level, Re-order Quantity, Manufacturer Code.

(b) Customer table, Order table.

3. (a) (See Chapter 12 (1).

(b) Membership number needed to uniquely identify the member.

4. Data independence: structure does not affect the programs which access the data so set up time for new applications is reduced; quality of management information is improved; increased productivity as ad hoc reports can be generated to meet particular needs;

consistency of data: less data duplication so errors due to discrepancies are reduced; input preparation reduced to 'single input' principle; **control over redundancy**: updating less time-consuming as data duplication is minimised; less storage space required; integrity of data as DBMS can specify constraints when data is added; greater security of data.

5. See chapter.

6. (a) See chapter

(b) Customer (<u>CustId</u>, Surname, Forename, Street, Town, Postcode, Telephone)
Account Information (<u>AccountType</u>, Interest Rate)
CustAcc (<u>AccountNum</u>, CustId, AccountType, Frequency, LastStatement, Balance)
Transaction (<u>TransNum</u>, AccountNum, Date, Amount)

(c) Use query to find all accounts with statements due on chosen date using a calculation to work out date from date of last statement and frequency; find all transactions for these accounts; list transactions in date order.

7. (a) Query by example, structure query language.

   (b) QBE: easier for novice – standard grid but limited options; SQL: greater flexibility in criteria but takes longer to learn skills and greater chance of error.

8. (a) Book(<u>BookNum</u>, Title, Category); Member(<u>MemberNum</u>, Name, Phone, …)
   Loan(<u>BookNum</u>, <u>MemberNum</u>, <u>LoanDate</u>)

   (b) BookNum field links Book table to Loan table; MemberNum field links Member table to Loan table.

9. (a) see Q8 but replace book with video.

   (b) See text.

10. (a), (b), (c) See text.

11. ***** part a. in the question should be part of the explanation. Part b c and d should be called a b and c.
    a. See text.
    b. 4 tables could be used:
    Villa (<u>VillaNum</u>, Name, address, Resort, Weekly cost, OwnerNum, etc): *VillaNum is primary key*
    Customer (<u>CustNum</u>, Name, Address, etc) *with CustNum the primary key*
    Booking (CustNum, <u>VillaNum, date</u> etc) *with VillaNum and Date the composite primary key*
    Owner (OwnerNum, Name, Address etc) *with OwnerNum the primary key*
    VillaNum links booking to Villa, OwnerNum links owner to Villa, CustNum links Customer to booking.

Chapter 14

1. Package does not support WYSIWYG. Wrong printer driver installed.

2. A printer driver is the software that configures the computer for a particular printer. It is necessary because different printers operate in different ways (different speeds, line-lengths, etc.).

3. Provides interface between the operating system and the printer; translates formatting and highlighting information into a form that the printer can understand; stores page set-up; translates fonts, bit maps and size control.

4. (a) Manages all the other programs in a computer; manages user communication with the computer; handles input/output from attached hardware devices; manages memory, backing storage.

   (b) An applications package that is appropriate to many areas of day to day business operations. Word-processors/DTP, spreadsheets, database management systems, integrated.

    (c) Allows the user to specify the interface and the functions required, then automatically generates the code to produce the customised application.

5. See text.

6. See chapter.

7. See chapter.

8. See Q 5.

9. e.g. copy and paste, find and replace, undo etc. Note that formatting facilities such as changing font, colour, size, bold print, etc are editing not facilities and are not acceptable answers.

Chapter 15

1. User B is likely to be using a later version of the software, which is not backwards compatible. User A is using an earlier version and so cannot read files created by User B in the later version. User A needs to either upgrade to the later software version or install a utility that will convert User B's files into the earlier format.

2. The software may not function correctly if the specification of the PC is too low, e.g. it does not contain enough RAM. The software may not have been fully tested on every possible hardware platform due to commercial pressures. The systems software may be incompatible e.g. a different version of Microsoft Windows.

3. (a) The ability to store several commands as one named command.

    (b) When performing any common task, e.g. putting an address at the top of a letter, clearing data from a spreadsheet.

4. (a) Files created in one package can be imported into another package.

    (b) It avoids retyping data which can be time consuming, expensive and lead to mistakes.

5. (a) Used to generate part of a complete system, set up automatically.

    (b) Can generate reports from data in files independently of the content, can perform calculations.

    (c) See pages 174 and 175.

6. Data files created under the old version of the software may not be usable in the new version; Staff may need to undergo training to use the new version; the hardware used may not be fast enough or have adequate memory to run the new version.

7. Difficulties include commercial pressures to publish quickly, the large number of different PC platforms available, the complicated nature of software with thousands if not millions of possible paths to test. Measures to minimise these problems may include better in-house testing and getting users to test on their PC (beta testing).

Chapter 16

1. Payroll, producing statements for customers, updating standing orders.

2. See chapter.

3. (a) Faster response to enquiries, information more up to date.

   (b) Staff may not be confident so make more mistakes and be slower; data could be lost if unexpected errors occur.

4. When many changes need to be made to a file of data at the same time, when cost of collecting data in real time would be too much.

5. See chapter.

6. See chapter.

7. See chapter.

8. See chapter.

9. Date, sound, picture or graphics, program instruction, real numbers or integers.

10. (a) All data is collected together over a set period of time into batches of a set size to be processed in one computer run without any human intervention.

    (b) (i) The processing can be done when computer system is quiet, it requires less staff and less hardware
        (ii) Not all students details will be updated if forms are incorrectly completed or lost. Details may be out of date for up to a week and error corrections may take further week.

    (c) Transaction processing deals with each set of data as it is submitted. Each transaction is completed before the next is begun.

Chapter 17

1. (a) The ability to produce several slides and to move between slides.

       The ability to reveal part of a slide at a time. (These answers must be major functional features – including sound or animation is not a major feature.)

   (b) Visual using LCD projector, oral using sounds from a loud speaker.

   (c) People at the back of the room may not be able to see, in which case a larger image should be used or a larger font should be used for text.

2. Information can be presented in text, table or graph form. This capability is needed so that the presentation method is suitable for he particular audience? Do they want detail? Do they want a summary? Do they need to see trends at a glance? Do they need to compare performance (e.g. with last year)?

3. (a) Can include sound, can include clip-art, can have animation, can edit it easily, can have automatic timing…

   (b) Distance of audience from screen affect size of text, colour scheme used – must be good contrast, do you want to reveal the whole slide at once or just part?

4. Results can be presented in text, table or graph form depending on the needs of the governors.

5. (a) e.g. physical security, passwords, good practice.

   (b) (i) Sound, animation, automatic timing…
       (ii) Size of text, contrast.

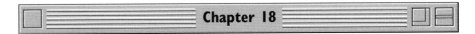
Chapter 18

1. See information in Chapter 18.

2. Computer memory has increased significantly and so can cope with sophisticated speech recognition software as well as other software, e.g. word-processing. Computers have got cheaper. Speech recognition techniques have become more reliable.

3. (a) A computer in each of the five shops can record sales. All the computers can be linked by a network, probably using public telephone lines. A modem and standard lines could be used or a faster ISDN line. As the data sent would only be sales data, very fast speed is not essential.

   (b) Details of sales could be input directly at the till, probably by scanning bar codes on the products. This sort of scanning is very common although other methods also exist, such as typing in the code number or reading a magnetic strip. Magnetic strips are also used on credit/debit cards used for payment.

   (c) Output devices at the till are in the form of a small printer to print the receipt and a display over the till to show prices, etc.

   (d) The likely storage devices would be internal hard disks. Today they can store over 20 Gigabytes and have very fast access times. These disks would store details of prices of products. When an item is scanned, the price is sent to the till so speed is important. The disks would also store details of today's sales for ordering and accounting purposes.

4. (a) Floppy disks store only a small amount of data. CD-ROMs store around 500 times as much. CD-ROMs cannot be rewritten to but floppy disks can. CD-ROMs have much faster access than floppy disks, …

   (b) (i) Transferring data between machines.
       (ii) Installing new software.

5. Writable CD. Relatively cheap. Can store large files.

6. (a) Sound files are too big for the floppy.

   (b) The sound files could be installation instructions.

Chapter 19

2. (a) To avoid permanent data loss and to ensure the integrity of stored data e.g. backup sufficiently up to date.

   (b) Selection of storage medium e.g. DAT Tape/RAID etc., selection of software with regard to facilities, recording of transactions, full or differential backup, frequency of copying, number of backup copies kept, recovery procedures, location of backup storage.

3. (a) Keyboard locks or swipecards to prevent use of equipment by unauthorised persons; floppy disk drive lock to prevent unauthorised copying of data removable disk drives so data can be kept away from equipment.

   (b) Password protection to prevent access to system, restriction of number of attempts to enter password, a firewall to prevent unauthorised remote access; a virus checker to prevent corruption of data; access levels to allow user a certain level of privilege e.g. read/write, browse, access to operating system etc.

   (c) Regular changes of password; prevention of poorly chosen password; staff not allowed to have written copies of password left by equipment to minimise chances of disclosure etc., careful vetting of users before they are allowed access rights.

4. a. b. see text

   (c) Customers might think they have made orders when the system is not working so these new orders are lost.

5. (a) All transactions during the week might be lost and they would need to be rekeyed in which would be time-consuming. Backup slow if more than one floppy disk is needed; floppy disks are unreliable.

   (b) The newsagent should use tapes rather than disks as they are more reliable and systematically label and organise them so that the appropriate tape can be found easily if required; a transaction log should be kept so that no transactions since the last backup are lost; back-ups should be kept away from the shop – ideally in a fireproof safe.

6. Security Procedures, e.g. not leaving terminals logged on; using appropriate back-up procedures; using encryption and virus checking; implemented levels of permitted access for users as well as passwords.

Chapter 20

1. Local Area Network – small geographic area, dedicated cables. Wide Area Network – wide geographic area, often using public telephone cables.

2. (a) Cheaper to install, no complicated network management required, relatively easy to set up from stand-alone systems, can share data easily.

   (b) Network interface card – to connect the computer to the cables. Network cables – to connect computers.

3. A gateway allows a Local Area Network to connect to a Wide Area Network.

4. (a) See question 1

   (b) Mention cost, maintenance, security, ease of installation and backup.

5. **Stand-alone** – cheap, poor security, must go back to the same machine to finish work, can't share printers easily. **Network** – good security, more expensive but can get network licences, can use any machine, can share printers, scanners, Internet access, can share data.

6. See question 5.

7. See question 5.

8. (a) Data can be shared without need for duplication; software can be shared; data transfer and communication is improved; centralised upgrading and installation of software; improved sharing of peripherals such as printers; central control of security and backup.

   (b) Central pool of data available to all employees; central control of security and backup; centralised upgrading and installation of software; dedicated servers usually provide faster access to network resources; users are freed from network management tasks; computers may be of different types.

   (c) Server; network Interface cards; transmission media; modem. Plus reasons for choice.

9. (a) Peer-to-peer or server based.

   (b) (i) Can e-mail his staff – better communication.
          Can share data – he has immediate access to booking details.
       (ii) Most than one member of staff can book at once – better service.
            Booking information more accurate – better service.

(c) Factors to be considered include: security of data, the need to transmit credit card details, whether to send confirmations by email, whether a permanent Internet connection is needed, type of connection to the net.

Chapter 21

1. Output needs to be designed in relation to different audiences which might include managers, customers, shareholders etc. Different methods of output may be used such as text, graphics or sound. Information might need to be detailed or aggregated.

2. (a) Windows, icons, menus, pointers.

   (b) This includes font types, font sizes bold, italics, underscore, superscript; no need for colour or special symbols to represent these effects; possible to see and manipulate the eventual layout on the screen more easily.

3. (a) Ease of transfer of skills to different packages; shorter learning time after learning first package; less training costs for employer; greater range of tasks accessible to users; confidence building for first time users.

   (b) Common tool bars, menu structure, functionality of pointing devices, icons; tool bars, buttons and icons in same place on screen; standard key combinations for short cuts; help in same format.

4. **Customers**: touch sensitive screen, use of colour; **agents**: Menu driven, text rather than graphics; **staff**: command drivem, more complex interface. All require justification.

   see Q3.

5. (a) Large tracker ball and customised keyboard for clumsy hands; simple, clear colourful display in bright, primary colours and very little text to keep child's interest.

   (b) For partially sighted large screen with large font; Braille printed output or voice output; voice input.

   (c) Large screen to display image, graphics tablet with icons to select and cross wire selecting device.

6. See text.

7. Technology is getting better – faster processors, more memory – so software is being developed that is more reliable.

8. See text.

9. (a) It is easier to learn new applications so the overheads for training are reduced and there would be an increase in productivity. Data can be easily transferred from one package to another.

(b) Common features include: commands to load or save documents, cutting and pasting commands, method of printing and previewing, command names, the order of menus and the grouping of submenus.

11. (a) There may be a touch screen where a screen button is touched to bring up a booking form.

(b) Interface could be menu driven where the user is led through a series of options.

(c) A screen (perhaps a touch screen) some sort of keyboard or pad and a printer would be required.

(d) Data that is entered should be verified and validated.

12. (a) Managing backing store, peripherals, privacy (passwords, etc), security (back-ups) etc.

(b) (i) Use of icons, menus, pointing device, windows.
(ii) Icons are easily remembered whereas it is difficult to remember commands, get the correct syntax as well as type them correctly.
(iii) Students only see those packages that they need to and are not confused by a cluttered screen. They are presented with the same screen on each computer that they use.

(c) To make sure that the new package can be installed and tested when there are no students accessing the network. To allow time for staff training and the development of teaching materials.

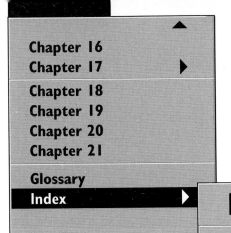

Chapter 16
Chapter 17
Chapter 18
Chapter 19
Chapter 20
Chapter 21
Glossary
Index

Index